Practical &
Decorative Concrete

Practical &
Decorative Concrete

Robert Wilde

Structures Publishing Co.
Farmington, Michigan

1977

Manufactured in the United States of America

Book edited by Shirley Horowitz

Book designed by Patrick Mullaly

Current printing (last digit)
10 9 8 7 6 5 4 3 2 1

Cover photographs: Round precast steps, upper left, courtesy of the Portland Cement Association, Skokie, Ill. Sculptured background, Paul Ritter "Sculpcrete," P.E.E.R. Institute, Perth, Western Australia. Hand float, center, Ann Arbor Photographics, Ann Arbor, Mich. Garden vases, lower right, Ann Arbor Photographics, Ann Arbor, Mich.

Library of Congress Cataloging in Publication Data

Wilde, Robert, 1923-
    Practical and decorative concrete.

    Bibliography: p.
    Includes index.
    1.   Concrete construction—Handbooks, manuals, etc.
I. Title.
TA682.4.W55   624'.1834   77-2833
ISBN 0-912336-38-2
ISBN 0-912336-39-0 pbk.

Contents

These irregular pieces, called "crazy paving," can be made either by molding your own, or using broken up pieces of pavement found lying unused.

Preface

Building with concrete is both an art and a science. This book can only be an introduction to some of the fundamentals—what concrete is, and how to build with it. The book illustrates some of the ways concrete is used around the home and garden, ranging from utilitarian to just-plain-fun projects.

These pages offer basic guidance for the person who may wish to do it himself, as well as for those who hire a contractor but want to know something about the material and the work involved. It summarizes knowledge collected from a variety of sources. Many craftsmen, manufacturers, contractors, and concrete engineers have offered information and illustrations; to all of them I express my thanks. In writing this book I referred often to publications and literature on concrete published worldwide. In that regard I especially wish to acknowledge the standards and recommended practices developed by the American Concrete Institute, and various publications issued by the Portland Cement Association and Concrete Construction Publications Inc.

—Robert Wilde

1. What Is Concrete?

The concrete you see, and walk on every day, is a mixture in which a paste of portland cement and water binds aggregates (such as sand and gravel or crushed stone) into a rocklike mass as the paste hardens through chemical action of the cement and water. The aggregate portion makes up about 60-75 percent of the concrete. This hardening process continues for years...concrete gets stronger as it gets older.

History

Man as builder has always sought a material that would bind stones into a solid, formed mass. The Assyrians and Babylonians used clay for this purpose. The Egyptians used lime and gypsum mortar as a binding agent in building the pyramids. The Greeks made further improvements and finally the Romans developed a cement which produced structures of great durability.

The early civilizations of Mesoamerica (Mexico and parts of Central America) were also using a mortar and stucco made from lime. In at least one region lime was used also as a kind of "concrete." It was poured into forms or given shape over an earth filling which was later removed.

The durable structures of the Romans are perhaps the best known to us today. The secret of Roman success in making cement was to combine slaked lime with pozzolana, a volcanic ash. This produced a cement capable of hardening under water. During the Middle Ages this art was lost, but in 1756 a British engineer, John Smeaton, discovered that cement made from limestone containing a considerable proportion of clay would harden under water...and the secret of hydraulic cement was rediscovered.

For some years large quantities of natural cement, produced by burning a naturally occurring mixture of lime and clay, were used. In 1824, Joseph Aspdin, an English bricklayer and mason, took out a patent on a hydraulic cement which he called "portland cement" because its color resembled the stone quarried on the Isle of Portland off the British coast. Aspdin's method involved the careful proportioning of limestone and clay, pulverizing them and burning the mixture, which was then ground into finished cement. Natural cement gave way to portland cement, a predictable product since its ingredients were controlled during the manufacture.

Portland cement today is a predetermined and carefully proportioned and ground combination of compounds of lime, silica, alumina, iron, and gypsum.

Portland Cement in the U. S.

The first recorded shipment of portland cement to the United States was in 1868, when European manufacturers began shipping cement as ballast in tramp steamers at very low freight rates. Portland cement was not manufactured in the United States until the 1870's. Probably the first plant to start production was that of David O. Saylor at Coplay, Pennsylvania. While Saylor was perfecting his product in Pennsylvania, Thomas Millen and his two sons were experimenting with the manufacture of portland cement in South Bend, Indiana. Their first portland cement was burned in a piece of sewer pipe and the resulting clinker (stony material fused together by heat and chemical action) ground in a coffee mill.

Today, of course, concrete is a major building material and portland cement readily available. One factor in the increase of cement production was the development of the rotary kiln. In the early days, vertical stationary kilns were used and wastefully allowed to cool after each burning. In 1885, an English engineer, F. Ransome, patented a slightly tilted horizontal kiln which could be rotated so that material moved gradually from one end to the other. This new type of kiln had much greater capacity and burned more thoroughly and uniformly, and rapidly displaced the old vertical kilns. Thomas A. Edison was a pioneer in the further development of rotary kilns. In 1902 he introduced the first long kilns used in the industry—150 feet in length. (Today's cement kilns vary from 300 to 700 feet long.) Parallel improvements in crushing and grinding equipment also influenced increases in cement production.

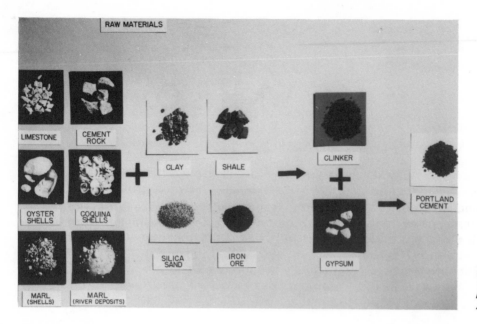

The raw materials that go into making portland cement (Portland Cement Assn.).

Concrete Components

Portland Cement

"Portland" cement is not a brand name. It is a type of hydraulic cement, just as "stainless" is a type of steel, and "sterling" a type of silver.

Portland cement, the basic ingredient of concrete, is a closely controlled chemical combination of calcium, silicon, aluminum, iron, and small amounts of other ingredients, to which gypsum is added in the final grinding process to regulate the setting time of the concrete. Lime and silica make up approximately 85 percent of the mass. Common among the materials used in portland cement are limestone, shells, and chalk, or marl combined with shale, clay, slate, or blast furnace slag, silica sand, and iron ore.

Portland cement manufacture requires some 80 separate and continuous operations, the use of a great deal of heavy machinery and equipment, and large amounts of heat and energy. Accompanying drawings show the steps in the manufacture of portland cement.

When rock is the principal raw material, the first step after quarrying is the primary crushing. Here pieces of rock, some as large as an oil drum, are reduced to about 6 inches. The rock then goes to secondary crushers or hammer mills for reduction to 3 inches or smaller.

The raw materials, properly proportioned, are then ground with water, thoroughly mixed, and fed into the kiln in the form of a slurry (a watery mixture of insoluble matter).

The raw material is then heated to approximately 2700 degrees F. in huge cylindrical steel rotary kilns lined with special firebrick. Kilns are 12 to 25 feet in diameter, large enough to hold an automobile, and are longer in many instances than the height of a 40-story building (300 to 700 feet). Kilns are mounted with the axis inclined slightly from the horizontal. The slurry is fed into the higher end; at the lower end is a roaring blast of flame.

As the material moves through the kiln, some elements are driven off in the form of gases. The remaining elements unite to form a new substance with new physical and chemical properties. The new substance, called "clinker", is formed in rough pieces about the size of marbles or as large as golf balls.

Clinker is discharged red hot from the lower end of the kiln and brought down to handling temperature in various types of coolers. (The heated air from the coolers is returned to the kilns, saving fuel and increasing burning efficiency.)

The clinker can be stockpiled or conveyed to a series of grinding machines. Here gypsum is added and the cycle is completed. The final grinding operation reduces the clinker to a fine powder.

Most cement is shipped in bulk to large users such as ready-mixed concrete plants or precast concrete and block plants. Cement in smaller quantities is shipped in strong paper bags (94 pounds), to become available at your local building supply center.

American Society for Testing and Materials (ASTM) specifications provide for five types of portland cement. Type I is the most widely produced and is used for general construction purposes. Types II through V are used for special conditions and would seldom be used by the homeowner for normal projects.

There are also some special purpose cements. One, white cement, is worth mentioning. Normal portland cement is grayish in color. White cement is produced from raw materials low in iron, and makes concrete with the same structural strength as gray cement concrete. White cement is used for architectural work, especially with certain colored aggregates. It is also popular where brilliance is desired, for projects where light-reflectance is important, and for colored concrete. With the addition of pigments, white cement produces very light shades of pastels and other colors which are not possible with gray cement. In come areas of the country rainfall will keep white cement concrete clean, and a wash job, if required, will make it look like new.

Aggregates

The aggregates used in residential concrete work are mainly sand and gravel, crushed stone, and occasionally blast furnace slag. There are lightweight aggregates (vermiculite, perlite, and expanded shale), as well as special heavyweight aggregates, but the homeowner is not likely to use these in his projects.

Aggregates make up the major volume of concrete, about 75 percent of its volume. Both the cost and the quality of the concrete are affected by the kind of aggregates selected. Aggregates greatly affect the amount of cement necessary to produce strength and other concrete properties. The size, shape, texture, and grading of aggregates also affect the workability and ease of placing and finishing concrete.

Fine aggregate consists of sand or other suitable fine material. A good concrete sand will contain particles varying uniformly in size from very fine dust-size up to ¼-inch diameter. Good gradation from coarse to fine is important for the workability of the concrete.

Coarse aggregate consists of gravel, crushed stone, or other suitable materials larger than ¼ inch. Coarse aggregates that are sound, hard, and durable are best for making concrete.

All aggregates, fine and coarse, should be clean and free of dirt, loam, clay, or vegetable matter such as roots and leaves; these foreign particles prevent the cement paste from properly binding the aggregate particles together, thus producing porous and low-strength concrete.

The largest size of coarse aggregate you can use will depend on the kind of work for which the concrete is to be used, varying from ¼ inch up to a maximum of 1½ inches. In walls, the largest pieces of aggregate should not be more than one-fifth the thickness of the finished wall section. In slabs for driveways, sidewalks, and patios the maximum aggregate size should be approximately one-third the thickness of the slab. Accordingly, 4-inch slabs may use coarse aggregate with 1-inch maximum size and 5- to 6-inch thick slabs could use 1½-inch maximum size aggregate. The largest pieces of aggregate should never be larger than three-quarters of the width of the narrowest space through which the concrete will be required to pass during placing. The narrowest space is usually the space between reinforcing bars or between the bars and the forms.

This is how a well-graded coarse aggregate looks. Well-graded aggregates have particles of various sizes with the smaller pieces fitting among the larger ones, and no excess of any one size. The aggregates shown here vary from ¼ to 1½ inches. (Portland Cement Assn.)

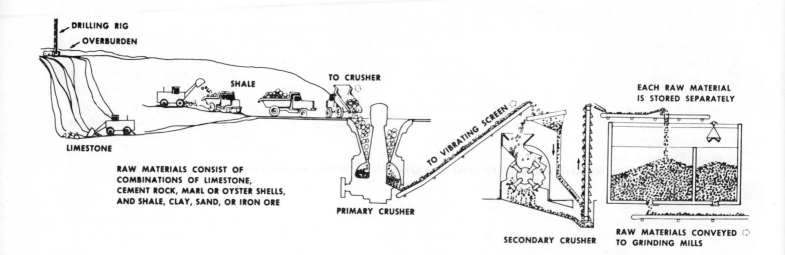

1 Stone is first reduced to 5-in. size, then ¾ in., and stored

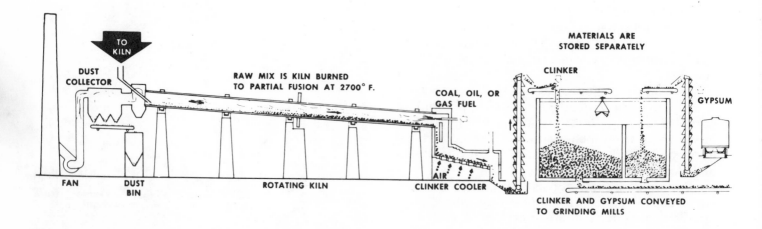

3 Burning changes raw mix chemically into cement clinker

A flow chart of typical steps in the manufacture of portland cement (Portland Cement Assn.).

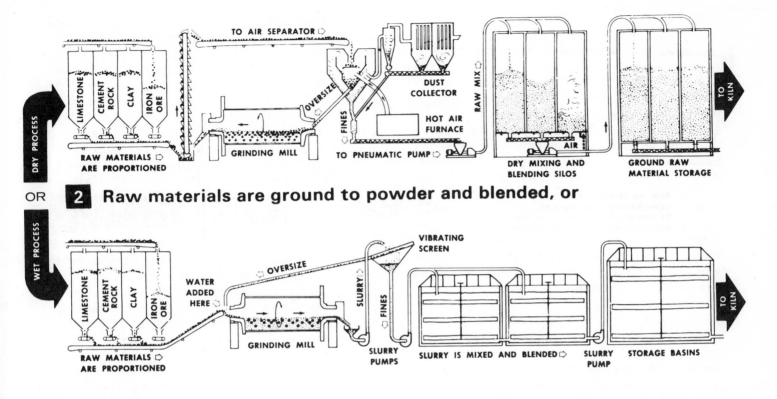

2 Raw materials are ground to powder and blended, or

2 Raw materials are ground, mixed with water to form slurry, and blended

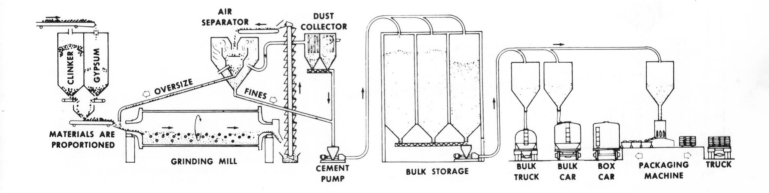

4 Clinker with gypsum added is ground into portland cement and shipped

Water

Water for making concrete should be clean—not muddy and full of silt—and free of oils, acids, algae, and sewage and industrial waste. In general, water that is fit to drink is suitable for making concrete.

Sea water is suitable for mixing concrete, although it offers some problems. Sea water has a concentration of about 3.5 percent salts. It makes concrete of higher-than-normal early strength (early setting time is reduced) but the strength is reduced at later periods. If the concrete is reinforced, the salts increase the corrosion potential for the reinforcing steel. Sea water increases surface dampness and efflorescence, so it should not be used where surface appearance is important or where concrete is to be painted or plastered.

Admixtures

An admixture is a material other than aggregates, portland cement, or water that is used as a concrete ingredient and is added to the batch immediately before or during its mixing. Admixtures are used in concrete for a variety of purposes, such as to improve workability, reduce segregation, entrain air, or to accelerate or retard setting and hardening.

The homeowner is likely to use only two kinds of admixtures: retarders and air-entraining agents, although there are many more types.

The principal use for admixtures having a retarding effect on the set of cement in concrete is to overcome the accelerating effect that temperature has on setting during hot weather concreting, and to delay early stiffening action of concrete. However, the homeowner who wants to retard the set of a surface layer of mortar so that it can be readily removed by brushing, thus exposing the aggregate and producing a textured surface, is more likely to use a retarder solution applied directly to the surface of the concrete.

Air-entraining admixtures used in concrete improve the workability and durability. For exposed concrete in sidewalks and driveways, air entrainment produces a concrete resistant to severe frost action and to the effects of salt used to melt ice and snow. If concrete is to be exposed to freezing and thawing, it should be air-entrained concrete. It is easy to buy air-entrained concrete from a ready-mixed concrete plant, and not too difficult to add an air-entraining agent to concrete being mixed in a mixer at the job site. However, for very small jobs mixed by hand, hand-mixing will not be vigorous enough to make air-entrained concrete.

2. Concrete and Its Uses

Although most people recognize the more common uses of concrete around the home and for the paving of streets and highways, few realize how widely it is used and how dependent we are on its versatility. Concrete is indispensable for: sidewalks, highways, airport runways, bridges, and dams; in the construction of large and small buildings; in harbors and offshore structures; as well as a multitude of other major and minor projects. Both farmers and city dwellers use it in innumerable ways.

Early Pavements

America's first concrete pavement was built in Bellefontaine, Ohio, in 1891. The most extensive early use of concrete for street paving was in Canada; in 1907, more than 2 miles were laid in Windsor, Ontario. The first mile of concrete road in the United States was built in 1909 in Wayne County (Detroit), Michigan.

Reinforced Concrete

Reinforced concrete was first used in 1850 when Joseph Monier built thin-walled concrete tubs, tanks, and garden pots with metal reinforcement. The French gardener was granted his first patent in 1857. Another Frenchman, Joseph Louis Lambot, built a reinforced concrete boat, which he exhibited at the Paris exposition in 1855.

In 1870, William Ward, a mechanical engineer living in Port Chester, New York, experimented with reinforced concrete in order to construct a fireproof house. Completed in 1875, the Ward House is acknowledged as the first reinforced concrete building in the United States. The interior and exterior walls, cornices, and towers were constructed of concrete. Beams, floors, and roofs were made of concrete reinforced with light iron beams and rods. The house is still lived in today, an outstanding example of the durability of cast-in-place concrete.

Important practical applications of reinforced concrete for structures were developed by E.L. Ransome, working on the Pacific Coast. These and other

The "first" reinforced concrete building in the United States was this house built in 1875 by William Ward in Port Chester, N.Y. (New York Concrete Construction Institute)

applications dating back to the last quarter of the 19th century established a style of construction that was both simple and practical.

The rapid adoption of reinforced concrete was based on the fundamental knowledge that concrete is stronger in compression than in tension. When concrete structural members must resist tensile stresses, steel supplies the necessary resistance to pulling apart. Reinforcement can be either steel bars or mesh embedded in the concrete.

Architectural Concrete

Late in the 19th century, concrete was usually covered with gingerbread veneers. Although it was openly welcomed as a structural boon, few people recognized its ornamental potential.

But today there is an infinite variety of shapes and textures in concrete: sculptured concrete

A new kind of architectural art form in concrete from Australia called Sculpcrete. These concrete walls were cast in an expanded polystyrene formwork lining that was carved with special tools. (Courtesy Paul Ritter, Humanizing Concrete, PEER Institute Press, Perth, Western Australia, 1976)

panels, walls—either precast or cast-in-place—in color set off with exposed aggregates, precast grilles and screens, concrete surfaces smooth as glass or rough as broken stone. Fluting, rustications, relief patterns, and other ornamental devices are executed easily. Concrete offers a wide range of decorative possibilities since it is plastic in its initial stages, reproduces the pattern and texture of the mold or formwork into which it is cast. It can also be treated in a variety of ways once in its hardened state to reveal color, particle shape, and surface texture of the aggregates in the concrete.

The appearance of concrete depends mainly on three factors: color, texture, and pattern. Both the cement and aggregate contribute to the color of the concrete. The appearance of concrete owes as much to texture as to color, and the variety of textures available is broad. Pattern can be obtained in a variety of ways and will be affected by treatment of joints, shape, arrangement, and use of cast or "profiled" areas.

Many pleasing decorative finishes can be built into concrete during construction. Color may be added to the concrete through use of white cement,

Exposed aggregate finishes are not only attractive, but also resistant to wear and weather. (Portland Cement Assn.)

Almost any geometric design can be stamped or scored into a concrete slab. This driveway was scored with a simulated random flagstone pattern. (Portland Cement Assn.)

pigments, and exposure of colorful aggregates. Textured finishes can vary from a ground smooth polished surface to a bush-hammered rugged texture. Geometric patterns can be scored or stamped into the concrete surface.

Concrete Masonry

The term "concrete masonry" is applied to block and brick building units molded of concrete and used in all types of masonry construction and in landscaping. Shown here are a few of many shapes and sizes for conventional wall construction. There are concrete brick and slump and split units (see illustrations). The chimney and column units shown have also been used to build screen grilles, along with the screen wall units and architectural concrete masonry units (see illustrations).

The great variety and architectural flexibility of concrete masonry make it adaptable for use indoors and outdoors. It can be used in all-concrete construction or in combination with other materials.

Concrete masonry can be made colorful by painting or by using integrally colored units.

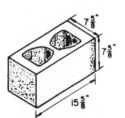

Regular   Frog   Double

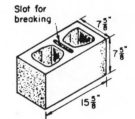

Slump   Split

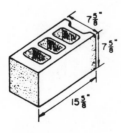

Hollow-perforated

Some sizes and shapes of concrete brick. (Portland Cement Assn.)

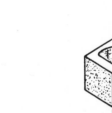

Regular stretcher

One plain end (single corner)

Both ends plain (double corner or pier)

Slot for breaking

Two-core 8 x 8 x 16-in. units

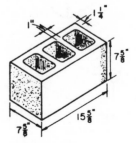

Regular stretcher

One plain end (single corner)

Both ends plain (double corner or pier)

Three-core 8 x 8 x 16-in. units

Conventional concrete masonry wall construction uses either two- or three-core units. The nominal block size is 8 x 8 x 16 inches. (Portland Cement Assn.)

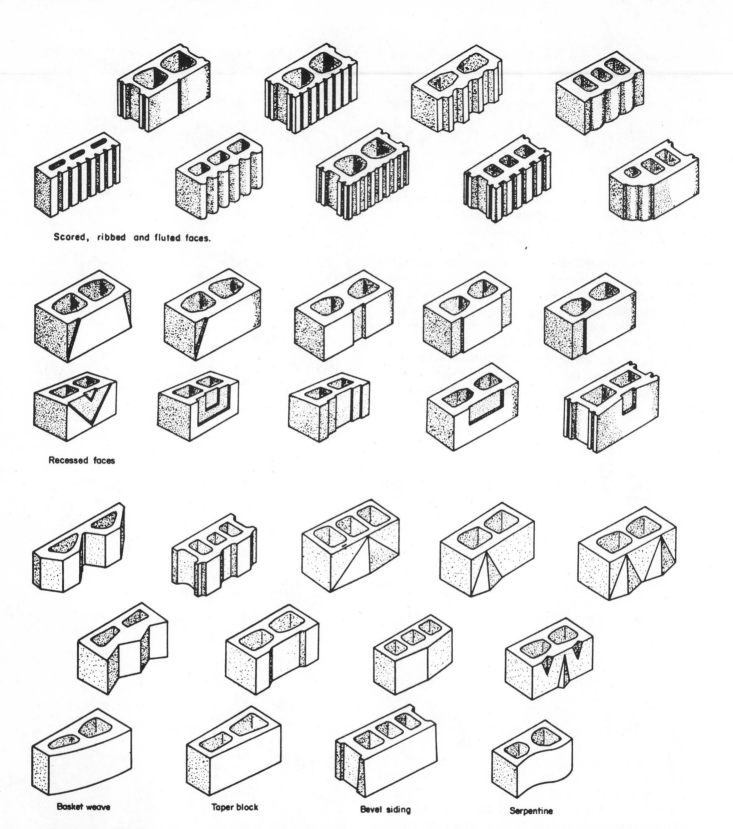

Scored, ribbed and fluted faces.

Recessed faces

Basket weave   Taper block   Bevel siding   Serpentine

Angular and curved faces

Architectural concrete masonry units. Patterns molded in the block machine include scored, ribbed, and fluted faces; recessed faces; molded angles and curves. (Portland Cement Assn.)

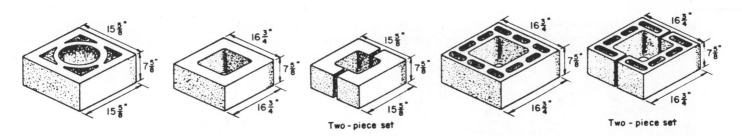

For 8-in. rounded liners   For 8-in. square liners

Two - piece set   Two - piece set

Chimney and column masonry units. These utilitarian units can also be used to build screen walls. (Portland Cement Assn.)

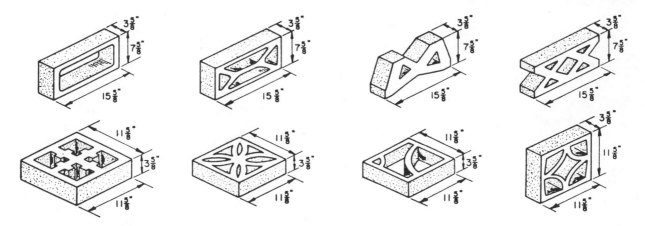

Screen wall units, sometimes called screen block or grille block, are both decorative and functional. They are used in room dividers, partitions, garden fences, and patio screens. (Portland Cement Assn.)

Integrally colored or painted, rough-textured split brick or block provide tasteful accents for exterior finish. Various sizes, such as 4 x 2 x 16 inches and 4 x 4 x 16 inches, are available. Split block is produced by splitting or fracturing concrete masonry units lengthwise.

Variety may be added through the use of attractive slump block with tooled mortar joints, and through variations in the color, size, and pattern of the block. Slump block are appropriately named. The concrete mix used has a consistency that, when the units are released from their molds before complete setting, sag or slump. The slightly irregular blocks range in size from 1⅝ to 3⅝ inches high.

Sculptured or "shadow" 8 x 8 x 16-inch block are available in a variety of surface treatments. These units allow for a great variety of geometric designs and create interesting light and shadow patterns.

An open garden-patio fence built with concrete split brick. (National Concrete Masonry Assn.)

A flower planter of split block. Solid concrete masonry units are split mechanically to reveal aggregates along with an attractive rough texture. (National Concrete Masonry Assn.)

Concrete masonry units are usually dimensioned in modules of 4 or 8 inches. In masonry construction the ⅜-inch thick mortar joint has become standard. Thus, the actual units are molded to account for the mortar joint, so that when placed in the wall the lengths, heights, and thicknesses are in multiples of the 4- or 8-inch module. Thus the actual dimension of an 8 x 8 x 16-inch block (the most widely used size) is 7⅝ x 7⅝ x 15⅝ inches.

Mortar joints should be maintained as closely as possible to the ⅜-inch dimension, and should not exceed ½ inch. In general, the thinner the mortar joint, the stronger the wall.

Ready-Mixed Concrete

Ready-mixed concrete can be made in a central mixing plant and hauled to the job site, or its components can be measured at a central proportioning plant and mixed in transit or at the job site. It is delivered to the consumer in a plastic and unhardened condition.

A ready-mixed concrete plant in some ways resembles the operation of a custom tailor, where every suit is made to fit the customer's specifications. Similarly, in a ready-mixed concrete plant each batch of concrete is tailor-made to the customer's specifications and requirements.

Except for the smallest job, usually the most convenient and economical place for the homeowner to buy concrete is the ready-mixed concrete producer. There are more than 6000 ready-mixed concrete producers in the United States. The ready-mix producer handles the job of proportioning, weighing, mixing, and delivering, and can supply concrete to meet the requirements of any project.

Slump brick, with its adobe profile, makes an attractive fence. (National Concrete Masonry Assn.)

One style of architectural concrete masonry unit is called "Shadowal." Geometric patterns can be created in the wall, enhanced by the play of light and shade on the profile faces (National Concrete Masonry Assn.).

Shadow row combines regular block with a raised face block.

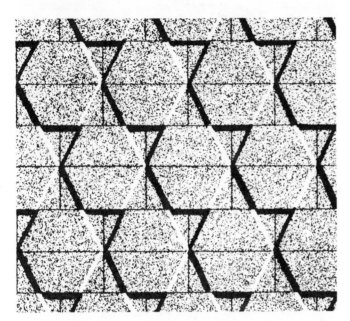

Filigree pattern has the appearance of hexagonal figures.

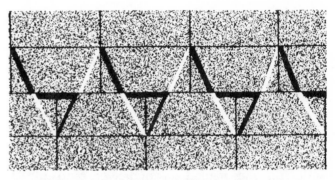

Pinnacles offer a striking example of combining the same raised face unit with regular block.

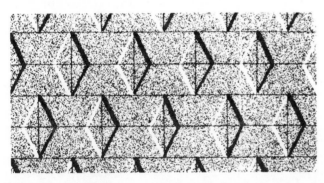

Waffle weave gives an overall diamond pattern.

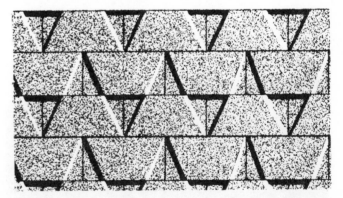

Fillip features raised face unit repeated throughout the wall.

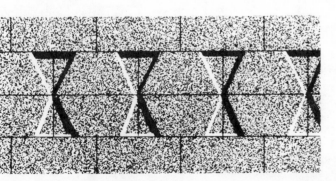

Fence design creates a border of hour-glasses.

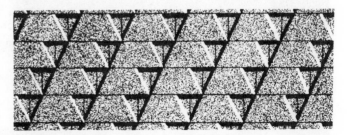

Sawtooth design again uses the raised face unit in a different way.

Ready mixed concrete can be delivered right to your driveway, sidewalk, or patio project. (Blaw-Knox Co.)

A driveway is not only functional, it can also be an attractive architectural feature. (Portland Cement Assn.)

Ready-mixed concrete is sold by the cubic yard (27 cubic feet) and most producers will deliver any quantity greater than 1 cubic yard (although some may set a 2 cubic yard limit). The cost of ready-mixed concrete varies with the type of mix, size of order, distance hauled, day of delivery, and unloading time. A call to one or more local producers will tell you what prices apply in your area.

Concrete Around the Home

Most homeowners prefer concrete sidewalks and driveways because of their durability, cleanliness, and smooth surfaces. They can be cleaned easily with a garden hose and their edges stay neat and trim. To provide a custom touch, some homeowners use exposed aggregate surfaces,

This fence of fluted, split-ribbed concrete masonry units laid in stacked bond offers both privacy and security. (National Concrete Masonry Assn.)

An attractive patio combination of screen block and exposed aggregate concrete slab. (National Concrete Masonry Assn.)

A short, step-down design of chimney block produces a lacy screen partition. (National Concrete Masonry Assn.)

squares, stripes, swirls, and scored designs for walks and drives.

Concrete is the ideal material for constructing footings and foundations because of its strength and resistance to underground destructive elements. If constructed on firm soil below the frost level, concrete footings will assure uniform distribution of the building weight and will eliminate any possibility of unequal building settlement. Concrete patios, barbecues, and garden walls add zest to outdoor living, and need practically no maintenance. Patios can be cast in place or constructed of colored patio block. Precast slabs may also be used for patios and walks. Paved play spaces are ideal for shuffleboard and basketball. A popular adjunct of modern outdoor living is the solar screen made with concrete block cast in special molds to give an "open" effect. This screen facilitates the passage of cooling breezes without obstructing scenic views or summer sunlight.

Whichever way you look at it, concrete serves many decorative and utilitarian purposes around the home and garden.

3. Some Fundamentals for Using Concrete

Knowing a little about concrete and its properties, and following a few basic principles of design and manufacture, will lead to good concrete. An understanding of these principles is helpful whether you do the concrete work yourself or hire a contractor.

Strengths and Weaknesses

Portland cement concrete is the most important masonry material used in modern construction. It is one of the most economical, versatile, and universally used building materials available.

Advantages

When first mixed, concrete is easily molded. In its plastic state concrete can be readily handled and placed in forms and cast into any desired shape. At this stage it may be troweled to a smooth surface, brushed to obtain rough texture, or shaped into ornamental patterns. It may be placed in specially made forms to reproduce intricate designs and sculptures. Pigments and colored aggregates may be added to provide a wide range of colors and textures. When hardened, concrete can be given a polished surface as smooth as glass or a chiseled, roughened texture. Concrete is as hard as rock, and especially strong under compression. Quality concrete work produces structures and products which are lasting, pleasing in appearance, and require comparatively little maintenance.

In the design of major structures and engineering works, the designer and constructor are concerned with many properties and characteristics of concrete. However, for projects around the home and garden, the properties of major importance are strength and durability. Appearance is also, of course, another important characteristic. Strength properties of special interest include concrete's strength in compression (its resistance to loads and forces which tend to crush it), and concrete's strength in tension (its resistance to loads which tend to stretch it). Durability involves the conditions to which the concrete is exposed, and include resistance to weathering, freezing and thawing, abrasion, drying, and wetting.

Disadvantages

Although well-made concrete is extremely strong in compression, it has little strength to resist bending. Recognition of the limitations of concrete will eliminate or minimize some of the structural weaknesses that can detract from its appearance and serviceability. The major limitations or disadvantages follow.

Low tensile strength. Concrete members which are subjected to tensile stress need to be reinforced with steel so that the tension will be taken by the steel, while the concrete carries the compression. Steel also tends to distribute shrinkage cracks in concrete; therefore, even concrete members which carry no tensile stress may contain steel to prevent the formation of large cracks.

The fact that concrete has low tensile strength also makes it important to build solid, rigid, and level bases under concrete slabs for floors and driveways. The accompanying illustration shows the importance of a solid base, as well as showing how concrete is weak in tension and cracks under such a stress. If the ground settles under the middle of a paving slab, a crack will start at the bottom when a load is applied. If the ground settles or washes out under the ends of a slab, the crack will start at the top when the load is applied. These cracks can be prevented if the base under the slab is properly compacted, level, and solid.

Drying shrinkage and moisture and temperature movements. Concrete, like all construction materials, contracts and expands under various conditions of moisture and temperature. This normal movement should be anticipated and provided for in the design and placement of joints as well as in the proper curing of the fresh concrete. Otherwise, damaging cracks will result.

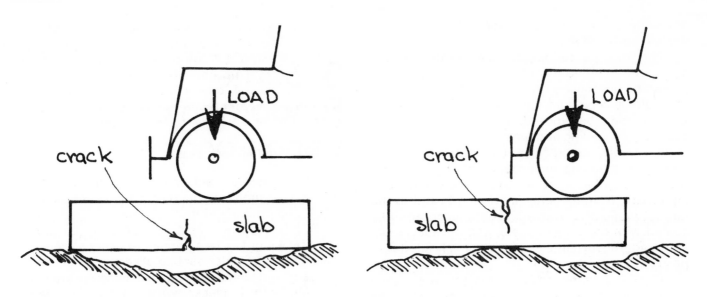

Cracks in slabs form under the load due to an uneven base or settlement of the ground.

While the concrete is setting and hardening, most of the mixing water will dry out, with the result that the concrete will shrink; the greater the amount of mixing water used, the greater the shrinkage will be. It is thus important to keep the concrete mix as "dry" as possible, rather than sloppy and soupy. It is better to tamp fresh concrete into a form than to pour it. The effects of shrinkage can be minimized by preventing the concrete from drying out too quickly, by keeping the concrete moist after placement through proper curing procedures.

Concrete will expand in hot weather and contract in cold weather. In hot weather concrete will expand about ½ inch in 100 feet, and if no precautions are taken uncontrolled cracking occurs. While there is no way to prevent movement of concrete with temperature changes, the cracking can be controlled through spacing of joints.

Also, reinforcement can be added to carry the tension due to the effects of temperature change or shrinkage. The amount of shrinkage and/or temperature reinforcement usually provided for is about one-fifth of 1 percent of the area of the cross-section of concrete.

Permeability. Even the best concrete is not entirely impervious to moisture. Concrete contains soluble compounds which may be leached out to varying degrees by water. Impermeability is particularly important where concrete is exposed to freezing and thawing, and also important in reinforced concrete where good dense concrete cover prevents rusting of the steel reinforcement. When the concrete will be frozen while wet, not only must a dense

cement paste be provided, but the concrete should also be protected with entrained air.

One of the major advances in concrete technology was the advent of air entrainment. The principal reason for using intentionally entrained air is to improve concrete's resistance to freezing and thawing exposure. Entrained air is also effective in preventing scaling of concrete surfaces caused by use of de-icing chemicals for snow and ice removal.

Useful Properties

Concrete is made up of aggregates (usually a mixture of fine and coarse) and paste (cement, water, air), with the aggregate particles in and separated by the paste. The concrete may contain an admixture to modify the properties of the fresh or hardened concrete. Concrete is one of the few building materials which exists in both a plastic and hardened state.

A plastic concrete is a concrete mix that is readily molded, yet changes its shape slowly if the mold is removed right after casting. The degree of

AIR 5%	WATER 15%	CEMENT 10%	AGGREGATE (FINE AND COARSE) 70 %
PASTE			MINERAL FILLER

Volumes of components of a typical concrete mix containing an air-entraining admixture (Reproduced courtesy Manual of Concrete Inspection, *American Concrete Institute, 1975).*

flexibility influences the quality and properties of the finished product.

The principal requirements of hardened concrete are: it should have the required strength; it should be homogeneous, watertight, and resistant to weather, wear, and other destructive agents to which it might be exposed; it should not shrink excessively on cooling or drying.

A durable, strong concrete is obtained by correctly proportioning and properly mixing the ingredients so that the entire surface of every particle of aggregate, from the smallest grain of sand to the largest piece of coarse aggregate, is completely coated with the cement paste, and the spaces between aggregate particles are completely filled with paste.

Quality concrete costs no more to make than poor concrete, but is far more economical in the long run because of its greater durability. The rules for making good concrete are simple:

1. Use proper ingredients.

2. Proportion the ingredients correctly.

3. Measure the ingredients accurately.

4. Mix the ingredients thoroughly.

Weight of Concrete

Concrete can weigh as little as 15 pounds per cubic foot or as much as 400 pounds per cubic foot.

By the use of heavy aggregates, dense concrete mixtures can be made that will weigh 250 to 400 pounds per cubic foot. With the use of special lightweight aggregates and foam, concrete weighing only 15 to 30 pounds per cubic foot can also be made—it will float, and can be sawed or nailed like lumber.

Normal sand-gravel concrete weighs about 155 pounds per cubic foot. This is the concrete which the homeowner will be working with, so care must be taken when molding precast units to not make pieces so large that they cannot be lifted and moved. For example, a precast patio slab measuring 2 x 1½ feet x 2 inches thick contains ½ cubic feet of concrete, and weighs around 78 pounds.

Strength of Concrete

The most common measure used to judge quality of concrete is compressive strength. The factors affecting strength usually also affect impermeability and resistance to weathering. Generally speaking, factors which affect compressive strength of concrete are: curing conditions, age of concrete, characteristics of cement, quantity of mixing water, quantity of cement, characteristics of the aggregates, time of mixing, and air entrainment.

Water-Cement Ratio

The most important single factor affecting compressive strength of concrete is the water-cement ratio—the quantity of mixing water per unit of cement. It is expressed as a ratio of the water to cement by weight or as a percent of water to cement (also gallons of water per bag of cement).

Sand and coarse aggregate in concrete are held together by the binder known as hardened portland cement paste. This paste hardens by the chemical reaction of portland cement and water.

The strength of the cement paste—and ultimately the durability, strength, and other properties of the concrete—depends on the amount of mixing water used. If too much water is used the paste becomes thin and diluted. When it hardens such a paste will be too weak to bind the aggregate firmly.

Although there are many other factors of concrete proportioning that affect strength—such as kind of aggregate, aggregate grading, type of cement, admixtures—the effect of the water-cement ratio is by far the most important. The water-cement ratio of concrete should be kept low in order to obtain high strength. The use of low water-cement ratio has other benefits also. It makes the paste more dense, and this means the cement paste will be less permeable, and the concrete more watertight. A strong, dense paste has less drying shrinkage and makes the concrete less susceptible to cracking.

In general, then, the more cement used per water (the richer the mix) the greater the strength of concrete. But cement content itself is not a measure of strength; rather, it is because the higher cement content gives a lower water-cement ratio with the quantity of water needed for workability. The lower the water-cement ratio, the stronger the concrete.

Recommended water-cement ratios for both air-entrained and non-air-entrained concrete are given in Chapter 4, "Making Concrete."

Air Content

When water freezes its volume increases by about 10 percent. If concrete is saturated with water and exposed to freezing temperatures, the water expansion may cause stresses greater than the concrete can resist. The first such exposure may cause some very small cracks, invisible to the eye. Repeated exposure to cycles of freezing and thawing can cause progressive cracking until visible cracks begin to develop and finally the concrete disrupts. The most common way to overcome the problem of low durability in freezing and thawing exposures is to include air in the concrete.

All concrete contains some air, usually less than 2 percent by volume. This entrapped air exists as scattered pockets about the size of the larger grains of sand. At times, however, we want to introduce additional air using an air-entraining agent or admixtures. These agents create a large number of spherical voids throughout the cement paste, about the size of the larger cement grains and finer sizes of sand.

This can be done easily and at virtually no cost by the inclusion of extremely small amounts of an air-entraining agent in the concrete mix, which produces microscopic air bubbles in the cement paste. During a period of freezing these bubbles serve as a safety chamber into which the freezing water can expand without developing pressures great enough to crack the paste.

Although air-entrained concrete was developed primarily to produce a concrete highly resistant to severe frost and cycles of wetting and drying or freezing and thawing—and thus to reduce scaling of paved surfaces—it also offered other side benefits. The air pockets in the paste make the fresh concrete more workable.

In most air-entrained concrete, 3 to 7 percent air is considered satisfactory; this is controlled by the amount of the air-entraining agent added to the mix. The air-entraining agents, which are chemicals especially made for this purpose, are added to the mixing water. The amount to be added to the mix depends on the brand of air-entraining agent. The ready-mixed concrete plant uses these agents to produce air-entrained concrete. Concretes with relatively small sizes of coarse aggregate contain more cement paste, and require more air than those with very large coarse aggregates.

There is another way to obtain air-entrained concrete: instead of buying the chemicals and adding it to the mix, many cement manufacturers market portland cements which contain an interground air-entraining agent. These cements are identified on the bag as "air-entraining." They can be bought at the same building supply center that sells regular portland cement.

Settlement and Bleeding

In newly placed non-air-entrained concrete, the solids will slowly settle through the body of water, usually leaving a layer of clear water known as bleeding. When entrained air is present, the rate of settlement is reduced, so that sometimes the bleeding water evaporates as rapidly as it appears. Air entrainment can either eliminate bleeding or reduce it by half, a desirable situation since bleeding delays proper finishing (floating and troweling operations should not be performed on a wet surface).

Curing

Through early curing the internal structure of the concrete is built up to provide strength and watertightness. The longer concrete is kept moist, the more durable and stronger it will become. In hot weather it should be kept moist for not less than 3 days.

Concrete is kept moist by a number of methods such as: water ponding; continuous sprinkling with water; covering with waterproof paper, plastic sheeting, or wet burlap; and, a liquid seal coat which hardens to form a thin protective membrane.

Curing and Strength Gain

The reaction of portland cement with water takes place slowly and continues indefinitely. As it progresses, the strength increases. The strength increases most rapidly during the first several days. It is important to properly cure concrete in order to gain the most strength.

When concrete is mixed it contains a larger amount of water than is actually needed for hydration of the cement content. It is nevertheless important to keep most of this water within the concrete during the curing period, so that the chemical reactions can be evenly distributed—resulting in stronger, denser paste.

Drying Shrinkage

After the curing period, unbound water evaporates. As water leaves, the microscopic interior surfaces of the cement paste begin to dry and develop a high tension that shrinks the concrete. All concrete undergoes this drying shrinkage. Typically a concrete slab made with crushed river gravel would have an ultimate drying shrinkage of about 0.06 percent. If this slab were a single 50-foot piece, it would shorten by ⅜ inch and create a gap averaging ³/₁₆ inch at each end. In actual practice the subgrade would sufficiently restrain the shortening and cause the slab to crack at intervals of 20 to 25 feet, creating openings roughly averaging ⅛-inch wide.

Shrinkage can be minimized by improvements in aggregate hardness or in the density of the paste. However, since shrinkage cannot be avoided entirely, it is good practice to take such shrinkage into account by strategically placing joints to control cracks.

What Happens if...

Improper technique	Possible consequence
Too high a water-cement ratio	Low strength Excessive bleeding Dusting and chalky surfaces Cracking Crazing Water leakage
Inadequate or improper curing	Low strength Dusting and chalky surfaces Curling Cracking Crazing Water leakage
Inadequate or improper jointing to accommodate drying shrinkage and thermal changes	Curling Cracking and spalling Water leakage
Inadequate air content in concrete exposed to severe weathering	De-icer scaling Cracking and spalling Complete disruption

Adapted from *Concrete Construction,* Nov. 1975.

---

4. Making Concrete

Most homeowners today buy ready-mixed concrete rather than mix concrete themselves. However, for those who rent a small mixer and mix their own concrete, it is well to review the basics of proportioning the ingredients. It is also good general information for the homeowner who buys from the ready-mix supplier.

Proportioning Concrete

Proportioning of ingredients is often referred to as "designing the mix." A properly designed mix will give desirable characteristics for both the fresh and hardened concrete.

The coarse aggregate provides the main bulk of the concrete. The sand is included to fill the spaces (voids) between the stones. The cement is required to bind the aggregates together in a solid mass, and so it must coat every particle, large and small, and fill the remaining tiny spaces in the sand. To produce good concrete, the materials must be carefully proportioned.

The principal factors to consider in proportioning are:

    a. requirements as to placing—location, size and shape of concrete job, amount of reinforcement;

    b. interrelationship of cement content, water-cement ratio, aggregate grading, and total water per unit volume;

    c. strength needed;

    d. quality of concrete needed to meet conditions of exposure;

    e. economy.

The size and shape of a concrete project and the amount and location of the reinforcement affect the consistency and workability of the concrete, and also the size of the aggregate. Consistency and workability depend on the relative quantities of cement, aggregates, and water, and on the grading of the aggregates.

By consistency, we mean the wetness of the mixture. There is no way to measure consistency, but the "slump test" can be very useful as an indication of consistency. The higher the slump, the wetter the mixture. "Plastic mix" is used to describe a consistency between the dry, crumbly consistencies, and the very watery consistencies. A "plastic mix" posesses cohesion and does not crumble; it flows sluggishly.

The term "workability" is used to describe the ease or difficulty which may be encountered in placing the concrete. It includes consistency and also where the concrete is to be placed. For example: a stiff, plastic mix with large aggregate which is workable in a large open form would not be placeable (workable) in a thin wall of a planter box.

Weight of Building Materials

Portland cement	94 pounds per bag
Masonry cement	70 to 85 pounds per bag
Water	8.33 pounds per gallon
Sand (dry)	97 to 117 pounds per cubic foot
Gravel	95 pounds per cubic foot
Crushed stone	100 pounds per cubic foot
Slag	65 to 70 pounds per cubic foot
Masonry mortar	100 to 110 pounds per cubic foot
Limestone concrete	145 to 155 pounds per cubic foot
Gravel concrete	145 to 155 pounds per cubic foot
Slag concrete	130 to 140 pounds per cubic foot

Qualities You Need in Your Concrete

The selection of concrete proportions involves a balance between reasonable economy, and the requirements of placeability, strength, durability, density, and appearance.

Strength. Strength is one of the basic characteristics or properties of concrete, and bears a fixed relationship to the the water-cement ratio. Differences in strength for a given water-cement ratio may result from changes in: maximum size of aggregate; grading of aggregate; texture, shape and strength of aggregate particles; and air content.

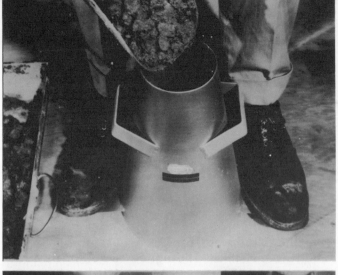

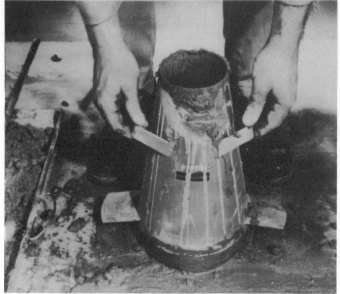

The slump test is a simple way to check concrete consistency. A special mold, called a slump cone, is 12 inches high, base diameter 8 inches, and top diameter 4 inches. The cone should rest on a firm, smooth nonabsorbent surface, such as a sheet of metal. Upper left: The cone should be held down by placing feet on projections attached to the base of the cone. Fill cone with freshly mixed concrete in three layers. Upper-right: Tamp each layer 25 times with a round steel rod. When filled, strike off the top.

Lower left: Lift the mold and allow concrete to settle. Lower right: Measure the slump by standing the cone next to the concrete, laying the tamping rod across the top of the cone, and measuring the distance from the rod to the top of the concrete pile. A small slump indicates a stiff consistency, and a very large slump, a very wet consistency. (Reproduced courtesy of Manual of Concrete Inspection, *American Concrete Institute, 1975)*

Durability. Concrete must be able to endure exposure to freezing and thawing, wetting and drying, heating and cooling, de-icing agents, and the like. Use of a low water-cement ratio gives a dense concrete which resists the penetration of moisture and aggressive liquids. Entrained air should be used in all exposed concrete in climates where freezing occurs. Resistance to severe weathering, especially freezing and thawing, and to salts used to melt ice is greatly improved through use of air-entrained concrete.

In ordering concrete or buying ready-mixed concrete, it is quite common to specify some or all of the following:

1. Maximum water-cement ratio;

2. Strength;

3. Minimum cement content;

4. Slump;

5. Air content;

6. Maximum size of aggregate.

Water-Cement Ratios for Strength. For the typical home project, your best compressive strengths (to be on the safe side) are somewhere between 3000 and 3500 pounds per square inch. This would call for water-cement ratios by weight of materials of about 0.68 for non-air-entrained concrete, and 0.59 for air-entrained concrete. This means that for a batch of concrete containing 10 pounds of cement and a water-cement ratio of 0.59 you would mix in 5.9 pounds of water. Water-cement ratio is also commonly specified in gallons of water per bag of cement. The equivalent of an 0.59 water-cement ratio is 6½ gallons of water per bag of cement.

Choosing Mixes for Small Jobs

On large building projects, trial mixes are made; this is seldom needed or economical on small backyard home projects. Usually you can use a more or less standard mix for small jobs or depend on your ready-mixed concrete supplier.

The preceding table is useful in simplifying mix design for small jobs. Follow this procedure and these proportions for a concrete that is strong and durable, as long as the amount of water added at the mixer is never large enough to make the concrete overwet.

Three mixes are given for each maximum size of coarse aggregate. For the selected size of coarse aggregate, Mix B is intended for initial use. If this mix proves to be oversanded, change to Mix C; if it is undersanded, change to Mix A. The mixes listed in the table are based on dry or surface-dry sand. If the sand is moist or wet, make the corrections in batch weight as spelled out in the footnote.

Workable Mix. *A workable mix contains the proper amount of cement paste, sand, and coarse aggregate. With light troweling, all spaces between coarse aggregate are filled with sand and cement paste. (Portland Cement Assn.)*

Concrete Mixes for Small Jobs

Procedure: Select the proper maximum size of aggregate. Use Mix B, adding just enough water to produce a workable consistency. If the concrete appears to be undersanded, change to Mix A and, if it appears oversanded, change to Mix C.

Maximum size of aggregate, in.	Mix designation	Cement	Sand* Air entrained concrete†	Sand* Concrete without air	Coarse aggregate Gravel or crushed stone	Coarse aggregate Iron blast furnace slag
½	A	25	48	51	54	47
	B	25	46	49	56	49
	C	25	44	47	58	51
¾	A	23	45	49	62	54
	B	23	43	47	64	56
	C	23	41	45	66	58
1	A	22	41	45	70	61
	B	22	39	43	72	63
	C	22	37	41	74	65
1½	A	20	41	45	75	65
	B	20	39	43	77	67
	C	20	37	41	79	69
2	A	19	40	45	79	69
	B	19	38	43	81	71
	C	19	36	41	83	72

*Weights are for dry sand. If damp sand is used, increase tabulated weights of sand 2 lb and, if very wet sand is used, 4 lb.

†Air-entrained concrete should be used in all structures which will be exposed to alternate cycles of freezing and thawing. Air entrainment can be obtained by the use of an air-entraining cement or by adding an air-entraining admixture. If an admixture is used, the amount recommended by the manufacturer will, in most cases, produce the desired air content.

Source: "Recommended Practice for Selecting Proportions for Normal or Heavyweight Concrete (ACI 211.1-74)," American Concrete Institute, 1974.

Wet Mix. *This mix is too wet; it contains too little sand and coarse aggregate for the amount of cement paste. (Portland Cement Assn.)*

Stiff Mix. *This mix is too stiff; it contains too much sand and coarse aggregate. (Portland Cement Assn.)*

Sandy Mix. *This mix is too sandy; it contains too much sand and not enough coarse aggregate. Although suitable for precast pottery or other small projects, it would not be economical for driveways or large jobs. (Portland Cement Assn.)*

Once the water-cement ratio has been established according to strength and durability requirements, the consistency needed for proper placement depends upon the maximum size of the aggregate, the size of the concrete project undertaken, and amount of reinforcement used. Typical slumps for concrete for home projects usually would vary from a 1-inch minimum to a 4-inch maximum (see photographs showing "slump test").

The table below shows recommended maximum size of aggregate for several types of construction and dimensions of section. The maximum size of aggregate should never be larger than one-fifth of the narrowest dimension between sides of the form, nor larger than three-fourths of the smallest clear spacing between reinforcing bars. In concrete slabs the best results come if the largest aggregate size is limited to one-fourth of the slab thickness.

Maximum Sizes of Aggregate Recommended for Various Types of Construction

Minimum dimension of section, inches	Maximum size of aggregate, inches		
	Reinforced walls, beams, and columns	Unreinforced walls	Lightly reinforced or unreinforced slabs
2½–5	½–¾	¾	¾–1½
6–11	¾–1½	1½	1½–3
12–29	1½–3	3	3

Source: *Concrete Primer*, American Concrete Institute, 1973.

The approximate cement content per cubic foot of concrete listed in the table is also helpful in estimating cement requirements for the job. The quantities in the table are based on concrete that has just enough water in it to permit ready working into

Stony Mix. *This mix is too stony; it contains too much coarse aggregate and not enough sand. (Portland Cement Assn.)*

forms without objectionable separation. Concrete should slide, not run, off a shovel. The table for mixes for small jobs is based on measuring your materials by weight. Measuring by weight rather than by volume is recommended because it is more accurate.

For small home jobs, however, it is sometimes neither possible nor convenient to measure the materials by weighing them. It is easy to measure by volume or parts. You can use pails, cans, or any sturdy box. An ordinary galvanized water pail makes a convenient batching container. The proportions given in the following table are by volume or parts. The mixes are similar, but not the same, when compared to the previous table giving mixes by weight.

Concrete Proportions by Volume

Maximum size coarse aggregate, in.	Air-entrained concrete				Concrete without air			
	Cement	Sand	Coarse aggregate	Water	Cement	Sand	Coarse aggregate	Water
⅜	1	2¼	1½	½	1	2½	1½	½
½	1	2¼	2	½	1	2½	2	½
¾	1	2¼	2½	½	1	2½	2½	½
1	1	2¼	2¾	½	1	2½	2¾	½
1½	1	2¼	3	½	1	2½	3	½

Source: *Cement Mason's Guide to Building Concrete Walks, Drives, Patios, and Steps,* Portland Cement Association, 1971.

Air Entrainment

Air-entraining admixtures added to the concrete, or use of air-entraining cement, will improve both the workability of fresh concrete and the durability of hardened concrete. For driveways, walks, and patios exposed to freezing and thawing and applications of salt for snow and ice removal, the use of air-entrained concrete is a must. Properly proportioned air-entrained concrete contains less water per cubic yard than non-air-entrained concrete of the same slump, and has better workability.

When protection from freezing and thawing is not required you may still wish to use air entrainment for its other advantages. Use air contents of about one-third less than shown in the table to reduce bleeding and segregation and improve workability.

The following percentages of air entrainment are recommended:

Maximum size aggregate, inches	Entrained air, %
⅜	7½ ± 1
½	7½ ± 1
¾	6 ± 1
1	6 ± 1
1½	5 ± 1

Remember...

Each of the principal ingredients of concrete—cement, aggregates, water—is indispensable in making concrete...but each can also negatively affect the properties of the wet or hardened concrete.

Quality. Cement and aggregates provide strength, durability, and volume stability to concrete. Too much water can destroy the quality of concrete, losing strength and durability.

Workability. Cement-and-water paste provides the workability for concrete. Too much aggregate makes the concrete unworkable and difficult to finish.

Economy. Aggregates and water are the least expensive materials in concrete. The use of too much cement makes concrete more expensive than it should be.

Measuring Ingredients

To produce good concrete, the materials must be carefully proportioned and measured accurately. As noted earlier, concrete ingredients may be measured by weight or by volume. Measurement by weight is best because it is more accurate and assures a greater uniformity from batch to batch. Also, it is easier to adjust mix proportions when measuring materials by weight.

The ready-mixed concrete supplier uses sophisticated equipment and computers to weigh and batch materials. For the home project, a common bathroom scale is accurate enough.

Each material should be weighed or batched in a separate container. Galvanized pails (2, 3, or 5 gallons) make good measuring units. Remember to "zero" the scale with the empty container on it. After weighing each material once, mark the level of the materials inside the container. Subsequent batches may be measured by using this mark. From this point on, check the weight occasionally, or when the moisture content of the sand changes.

Measurements made by volume are less accurate than by weight. Use boxes or buckets of convenient and equal size. Do not measure with a shovel. Measurement by shovelful is very inaccurate and little better than guesswork.

In proportioning by volume, fill each measure to the top and level it off. Measurement of the cement needs special care. Like all finely ground powders, cement bulks when filled into a container. To counteract this, the cement must be shaken down, usually by rapping the bucket or box sharply to settle the cement. Then fill the container up level with the top again.

An example of volume batching: for a 1:2¼:3 concrete mix, measure out 1 bucket of cement, 2¼ buckets of sand, 3 buckets of coarse aggregate, and ½ bucket of water.

Mixing Concrete

Proper mixing is an essential step in making good concrete. The materials must be uniformly distributed, and the surfaces of the fine and coarse aggregates well plastered with the cement-water paste. An even color and consistency in the concrete batch usually indicates thorough mixing.

Concrete can be mixed by hand or by machine. For very small jobs, where the volume of concrete is less than a few cubic feet, you may find it convenient to mix concrete by hand. Jobs small enough for hand-mixing can often be handled by using prepackaged concrete mixes sold in building material centers and hardware stores. Hand-mixing is not vigorous enough to make air-entrained concrete, regardless of whether air-entraining cement or an air-entraining agent is used.

If you are buying ready-mixed concrete, the supplier will follow prescribed procedures to furnish concrete that is adequately mixed and that fits your requested specifications.

Machine-Mixing

The prepackaged mixes are really only convenient for the very small job requiring only a few cubic feet of concrete; for jobs up to 1 cubic yard (27 cubic feet) it is probably more economical to buy the separate ingredients and mix them in a small mixer, when compared to cost of prepackaged mixes. For jobs requiring more than 1 cubic yard, ready-mixed concrete (if available) is without a doubt the most convenient way.

If you plan to mix your own concrete, the best way is to mix it with a concrete mixer. Small mixers from ½- to 6-cubic-foot capacity can be rented from a local rental service. To mix a 1-cubic-foot batch of concrete you will have to handle 140 to 150 pounds of materials (cement, sand, coarse aggregate, and water). The mixing time, after all ingredients are in the mixer, is generally 1½ minutes for mixers of 1-cubic-yard capacity or less, or until all materials are thoroughly mixed and the concrete has a uniform color.

Power concrete mixers, powered by gasoline or electricity, are available in several sizes and types. The electric-powered mixer is quiet and simple to use, but does require an electrical outlet. The gasoline-powered mixer can of course be operated anywhere. For most concrete jobs around the home, small mixers from ½- to 6-cubic-foot capacity can handle the job. While you can buy a mixer, it is usually cheaper to rent a mixer from your local rental service.

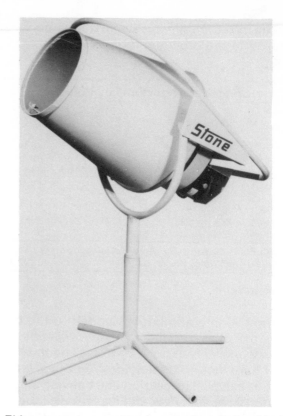

This compact concrete mixer can handle 2-cubic-foot batches and disassembles for transporting or storage. It is powered by an electric motor. (Stone Construction Equipment, Inc.)

The Rollamix is a real do-it-yourself concrete mixer— you furnish the power for mixing by strolling up and down the driveway. The dry capacity of the mixer is 100 pounds (about 12 average shovelfuls). (Decor Molds)

A recent invention combines manpower and machine mixing in a new way (see illustration). The mixer drum is molded of polyethylene and fitted with a rubber tire. The drum holds about 100 pounds of mix. The drum is filled with dry materials—cement, sand, coarse aggregates—and you stroll the mixer up the sidewalk to mix the dry materials. Add water, and keep on pushing the mixer for about 2 to 3 minutes. Then deliver the load to the work point and tip it to empty the concrete.

Rated capacity of a mixer is usually indicated on an identification plate fastened to the mixer. A mixer should never be loaded above its rated capacity. In selecting a mixer to rent, you should consider the size of your total concrete project...but keep in mind also how much concrete you want to handle in one batch for mixing and hauling.

Experts are not in complete agreement as to the best order of loading materials into the mixer. However, in general, you will get good results if you load ingredients into the mixer in this sequence:

1. With the mixer stopped, add all of the coarse aggregate, and half of the mixing water. (If a separate air-entraining agent is used, mix it with this part of the mixing water.)

2. Start the mixer, then add the sand, cement, and remaining water with the mixer running.

After all materials are in the mixer, continue mixing for at least 3 minutes, or until all materials are thoroughly mixed and the concrete has a uniform color. As the mix is discharged into the wheelbarrow watch to make sure it is "mushy" and workable, not "soupy." Concrete should be placed in the forms as soon as possible after mixing.

Adjusting the Mix

The first batch of concrete out of the mixer is also your "trial batch." The proportions of sand and coarse aggregates given in the tables earlier in this chapter are based on typical gravel aggregates. If these proportions do not give a workable mix with the aggregates you are using, you will have to adjust the mix.

As mentioned in the table "Concrete Mixes for Small Jobs," start with Mix B for the aggregate size you are using.

After mixing, put a sample amount of concrete into a wheelbarrow and check it for stiffness and workability. Use a shovel and a float or trowel to help you in judging the concrete. Work the concrete with a shovel: the concrete should slide down, not run off, the shovel. The concrete should be just wet enough to stick together without crumbling. Smoothing the sample lightly with a float or trowel also gives you a feel for workability and how the concrete will finish.

If the concrete sample is smooth, pliable, and workable, the concrete will place and finish well; this means the proportions used are all right and need no adjustment.

If this first trial batch is too wet, too stiff, too sandy, or too stony, you must adjust the aggregate proportions.

Again, refer to the table "Concrete Mixes for Small Jobs." If the initial mix is too sandy, change to Mix C; that is, decrease the amount of sand. If the initial mix is undersanded (too stony), change to Mix A; that is, decrease the amount of coarse aggregate.

A mix that is too wet contains too little sand and coarse aggregate for the amount of cement paste. Here is how you would adjust the mix. Weigh out about 5 to 10 percent more sand and coarse aggregate, depending on how wet the mix is. Add them to the trial batch in the mixer and mix for at least 1 minute. If the mix is still too wet, add some more aggregate and mix again until proper workability is achieved. Make sure you record the weight of the added sand and coarse aggregate. Now, in the next batch of concrete, use the original quantity of fine and coarse aggregate, but reduce the amount of water by 1 pound for every 10 pounds of aggregate you have added to the trial batch.

If the mix is too stiff, it contains too much aggregate. Most of the time this trial mix cannot be saved. **Never add water alone to a mix that is too stiff.** A trial batch that is too stiff to place might be saved by adding *both* cement and water in proportions of 1 pound of water to 2 pounds of cement. This increases the amount of cement paste and thus makes the concrete more workable. The adjustments for stiff mixes must be made in subsequent batches, by reducing the amounts of sand and coarse aggregates to give a workable mix. Record these new weights of aggregates and use that amount in all later batches.

Once the trial batch proportions have been adjusted, this mix can be used for the rest of the project as long as there are no changes in either the sand or coarse aggregate used.

Cleaning the Mixer

The mixer should be cleaned thoroughly each day it is in operation, or following each period of use before the concrete can harden. Clean out the mixer drum whenever it is necessary to shut down for more than 1½ hours.

By coating the outside of the mixer with form oil before starting, the cleaning job can be speeded up. Do not apply oil to the inside of the mixer drum. The outside of the mixer should be washed with a hose and all accumulated concrete knocked off. The thin cement film buildup on the outside of the mixer can also be removed with vinegar.

Hardened concrete should not be allowed to accumulate and build up in the mixer drum. To clean the inside of the drum, add water and several shovels (up to one-half the capacity of the mixer) of coarse aggregate while the drum is turning. Allow the drum to revolve for about 5 minutes. Discharge the aggregate and flush out the drum with water.

Concrete that has built up inside the mixer drum and has hardened can be removed by scraping and wire brushing. Do not use heavy hammers and chisels that might damage the drum and mixer blades. Do not pound the drum to remove hardened concrete; the concrete will just adhere more easily to the dents and bumps created. Hardened concrete can also be removed with a solution of hydrochloric acid (or muriatic acid): 1 part hydrochloric acid in 3 parts of water. Let the solution soak in for 30 minutes, scrape or wire brush, and rinse with water. Hydrochloric acid is hazardous and toxic, so be sure to follow adequate safety precautions.

Hand-Mixing

Although hand-mixing is less efficient than machine-mixing, it is often more convenient for small jobs, particularly if only a few cubic feet of concrete are required.

Hand-mixing should be done on a clean, hard surface such as a wooden platform, or on a clean, even paved surface. Or the concrete can be mixed in a mortar box or wheelbarrow. Concrete should not be mixed on the ground because it becomes contaminated with dirt and mud.

If a wooden platform is used, be sure the joints are tight to prevent loss of mortar, and that the plat-form is level. The wood surface should be moistened prior to mixing.

The measured quantity of sand is spread out evenly on the platform. The cement is spread evenly over the sand. Either a hoe or a square-pointed D-handle shovel can be used to mix the materials. Mix the cement and sand thoroughly by turning the two materials with the shovel or hoe until the color of the mixture is uniform. Streaks of brown and gray mean that the cement and sand have not been thoroughly mixed—keep mixing. Then spread this mixture evenly over the platform and dump the required quantity of coarse aggregate in a layer on top. Mix again until the coarse aggregate has been blended thoroughly with the cement and sand mixture. Turn the mixture at least three time; once it has been thoroughly mixed, form a depression or hollow in the center of the pile. Add the water slowly in the depression. Turn all the materials in toward the center and continue mixing until the water, cement, sand, and coarse aggregate are thoroughly mixed. Shovels should be kept down, close to the surface of the platform while turning the materials.

Packaged Concrete

For small projects such as patching, making small precast units, setting a clothesline pole, etc., you can buy prepared, prepackaged dry mixes. These come already proportioned, and need only the addition of water to the dry mix. Building material centers, and some hardware stores, sell a variety of packaged mixes. Sizes of packages vary, but common sizes are 45 and 90 pounds. For very small patch jobs, sand mixes are available in packages as small as 5 pounds. You can purchase a sand mix for patching purposes (or for casting small thin precast units like bookends), a mortar mix for laying brick and block or tuckpointing, or a concrete mix. A 90-pound package will make ²/₃ cubic feet of concrete. Directions for mixing and the correct amount of water to add are given on each bag. You must still do a thorough job of mixing to get good results.

5. Tools and How to Use Them

Required Tools

Tools used on the average concrete job include: (1) wheelbarrow, (2) shovel, (3) straightedge, (4) bullfloat or darby, (5) edger, (6) groover, (7) float, (8) trowel, (9) broom, and (10) water hose. One other should be mentioned even though it is not strictly a concrete tool—a tamper for use in preparing the subbase for flatwork (slabs). There are also tampers to compact concrete, but these are different tools.

Uses

Tamper

Preparation of the subgrade is the first step in building a sidewalk, driveway, or patio. This usually calls for a granular fill of sand, gravel, crushed stone or slag. This fill is compacted in layers. It should be tamped down to make it solid, with no air pockets.

Metal hand tampers are available. Any rather heavy object with a broad end may be used. A 4-foot piece of 4 x 4-inch timber makes a good tamper. Or a 4 x 4-inch block can be mounted on the end of a 3- or 4-foot-long dowel.

Shovel

Spreading and spading concrete for driveways and sidewalks are best done with a short-handled, square-end shovel. Ordinary garden rakes and hoes should not be used to spread concrete because they cause segregation (the separation of gravel or stone from the mortar in the mix).

Straightedge

A straightedge, also called a screed, is the first finishing tool used after the concrete is placed. It is used to strike off or remove the concrete in excess of the amount required to fill the forms and to bring the concrete surface to grade. A straightedge is often a

The granular subbase should be well compacted before concrete is placed for sidewalks, patios, and driveways. (Portland Cement Assn.)

Using a length of 2x4-inch lumber as a straightedge or screed, concrete is struck off by moving the straightedge back and forth with a saw-like motion. (Portland Cement Assn.)

straight piece of 2 x 4-inch or 1 x 4-inch lumber. Sometimes a ½ x 2-inch shoe strip is attached to the bottom. The most important thing is that the straightedge be straight and true. The straightedge should be longer than the widest distance between edge forms. Contractors sometimes use steel, magnesium, or aluminum straightedges.

Concrete is struck off by moving the straightedge back and forth with a sawlike motion. A small amount of concrete should be kept ahead of the straightedge to fill in low spots. About 30 inches should be covered in the first pass. As the straightedge is pulled forward, it should be tilted in the direction of travel to obtain a cutting edge. A second pass can be made if needed. During the second pass the straightedge should be tilted in the opposite direction.

Darby or Bullfloat

A darby is a long, flat, rectangular piece of wood, aluminum, or magnesium from 30 to 80 inches long and from 3 to 4 inches wide with a handle on top. It is used to float the surface of a concrete slab immediately after it has been leveled off. Darbying levels ridges and fills voids left by the straightedge. It also embeds the coarse aggregate slightly below the surface, in preparation for hand-floating and troweling.

A darby is used to float the surface of a concrete slab immediately after it has been screeded. Two types of wood darbies are shown. (Goldblatt Tool Company and Portland Cement Assn.)

A bull float is used to float large areas such as double driveways, which may be too large to reach with a darby. (Goldblatt Tool Company)

The darby should be held flat against the surface of the concrete and worked from right to left, or vice versa, with a sawing motion—cutting off bumps and filling low spots. When the surface is level, the darby should be tilted slightly and again moved from right to left, or vice versa, to fill any small holes left by the sawing motion.

A contractor or cement mason may use a bullfloat for the same purpose, which enables him to float a larger section of a wide slab.

Edger

Edgers come in many sizes; all are about 6 inches long, and vary from 1½ to 4 inches wide, with lips from ⅛ to 1½ inches. Stainless steel edgers with ½-inch radius are recommended for walks, drives, and patios.

Edging produces a neat, rounded edge that prevents chipping or damage, especially when forms

The working surface of the edger is flat with a lip on one side. (Goldblatt Tool Company)

are removed. Edging also compacts and hardens the concrete surface that is next to the form.

Delay preliminary edging until the concrete has set sufficiently to hold the shape of the edger tool. The edger should be held flat on the concrete surface. The front of the edger should be tilted up slightly when moving the tool in the forward direction. When moving the tool backward over the edge, tilt the rear of the tool slightly. You do not want to leave too deep a surface impression with the edger. In some cases, edging is required after each finishing operation.

The edger should be held flat on the concrete surface. Edging produces a neat, rounded edge. (Portland Cement Assn.)

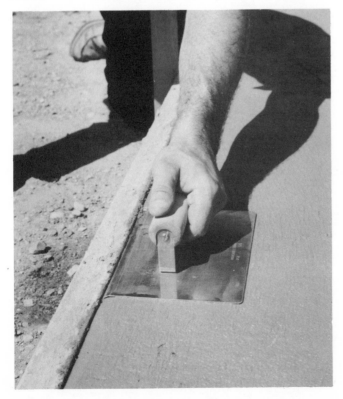

Jointer or Groover

During the edging operation, or immediately after, comes the jointing or grooving of a concrete slab. Proper jointing practices can eliminate unsightly random cracks. Most grooves are intended as contraction or control joints which predetermine and hide the location of cracks.

Jointers, sometimes called groovers, are made of stainless steel, bronze, or malleable iron in various sizes and styles. The tools are about 6 inches long and vary from 2 to 4½ inches wide, and have shallow, medium, or deep bits (cutting edges) running from $3/16$ to 1 inch deep. The cutting edge should be deep enough to cut the slab a minimum of one-fifth, and preferably one-fourth, of the depth. These joints, cut partly through a slab, will predetermine the location of possible cracks.

A jointer or groover is about 6 inches long, varies from 2 to 4½ inches wide, and the cutting edge (bit) for the joint can be shallow, medium, or deep. (Goldblatt Tool Company)

Cutting a control joint with a groover. A straight board makes a good guide for the groover. (Portland Cement Assn.)

It is good practice to mark the location of each joint with a string or chalk line on both side-forms and on the concrete surface. A straight 1-inch board at least 6 inches wide should be used to guide the groover. The board should rest on the side-forms, and be perpendicular to the edge of the slab. The groover should be held against the side of the board as it is moved across the slab.

To start the joint, the groover should be pushed into the concrete and moved forward while applying pressure to the back of the tool. After the joint is cut, the tool should be turned around and pulled back over the groove. This gives it a smooth finish.

Remember... if the concrete has stiffened to the point where the groover will not penetrate easily to the proper depth, a hand axe can be pushed into the concrete along the line. Then use the groover to finish the joint.

Hand Floats

The hand float is used to prepare the concrete surface for troweling. Floating follows edging and jointing, and has three purposes: (1) to embed large aggregate just below the surface; (2) to remove any imperfections left in the surface by previous operations; and (3) to compact the concrete and the mortar at the surface in preparation for troweling—or for that matter, to produce the finished surface.

Hand floats are made of aluminum, magnesium, or wood. Aluminum or magnesium floats usually come in two sizes, 12 to 16 inches long by 3½ inches wide. Wood floats are 12 to 18 inches long and 3½ or 4½ inches wide. Magnesium floats, recommended for most work, are light and strong, and slide easily over a concrete surface. Wood floats drag more; but the wood produces a rougher-textured concrete which serves well as the final finish when good skid-resistance is desired.

The hand float should be held flat on the concrete surface and moved with a slight sawing motion in a sweeping arc to fill in holes, cut off lumps, and smooth ridges.

Floating gives an even (but not smooth) texture. Since this texture has good skid resistance, floating is often the final finish. If so, it may be necessary to float the surface a second time after some hardening has taken place.

The surface marks left by edgers and groovers are removed during floating. If these marks are desired, for decorative purposes, the edger or groover can be rerun after the final floating.

Hand Trowel

Immediately after floating, the concrete surface can be troweled. Troweling produces a smooth, hard, dense surface.

Floating follows edging and jointing. The hand float should be held flat on the concrete surface and moved with a slight sawing motion in a sweeping arc. (Portland Cement Assn.)

Troweling produces a smooth, hard, dense surface. The trowel blade should be held flat on the surface, and moved across the surface in a sweeping arc. (Goldblatt Tool Company)

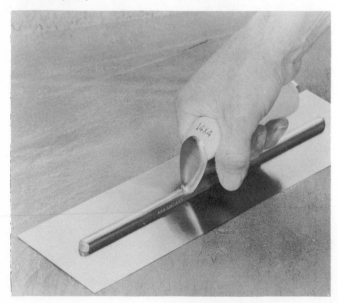

Can you tell where nature ends and man begins? These round patio slabs of exposed aggregate concrete blend right in with the natural stone path. (Portland Cement Assn.)

An "environmental wall" combining concrete patterns with natural plants. These pockets hold soil for planting. The wall was cast using the Sculpcrete process. (Reproduced from Humanizing Concrete, Paul Ritter, PEER Institute Press, Perth, Western Australia, Australia, 1976)

You can make almost anything with concrete. (Above left) This may be the biggest, boldest concrete frog you will ever see inhabiting a lily pool. (Above right) A whole family of turtles—all in concrete—add to the fun of this playground. (Below left) The exposed aggregate concrete pedestals for this picnic table lend a pleasing accent to the concrete patio. (Below right) Modernistic concrete flower pots fit well into this corner of a patio, with the open screen block fence as a backdrop. These projects can, for the most part, be created for only moderate cost and effort; their main ingredient (aside from concrete) is imagination! (Portland Cement Assn.)

Two types of fences for vastly different effects. (Below left) Integrally colored blocks give a brick effect. (Below right) Raked, rough texture offers both privacy and an unusual, attractive surface. (Portland Cement Assn.)

Sculptured concrete wall plaques with a delicate design coated with a colored glaze. Produced using the Sculpcrete process. (Paul Ritter, PEER Institute, Perth, Western Australia, Australia)

A fantastic variety of color and pattern is exhibited in this pottery and sculpture molded using the Sculpcrete process. The color and glaze are achieved by treating the polystyrene form linings with sealers and color and then dissolving the form lining. (Reproduced from Humanizing Concrete, Paul Ritter, PEER Institute Press, Perth, Western Australia, Australia, 1976)

A fence built with concrete posts and rails is virtually free from upkeep expense. Concrete fences cannot burn, rot, or rust.

Concrete slabs of two different finishes accent the elegance of this entryway courtyard. (Portland Cement Association)

This walkway between house and garage matches red brick with exposed aggregate concrete. (Portland Cement Association)

The steel hand trowel comes in many different sizes ranging from 10 to 20 inches long and 3 to 5 inches wide. Trowels are made from either carbon-tempered steel or stainless steel.

Normally at least two sizes of trowels are used. The first troweling usually takes a wide trowel 16x20 inches long. Shorter and narrower trowels are used for additional trowelings, as the concrete sets and becomes harder. For the final troweling, a 12x3-inch or smaller trowel, known as a fanning trowel, is preferred. If using only one trowel for the entire operation, it should measure about 14x4 inches.

For the first troweling, the trowel blade should be held flat on the surface. If it is tilted, ripples will be made that are difficult to remove later without tearing the surface. The hand trowel is used in a sweeping arc motion, each pass overlapping one-half of the previous pass. Each troweling thus covers the surface twice. The first troweling may be sufficient to produce a surface free of defects, but additional troweling may be used to increase smoothness and hardness.

There should be a lapse of time after the first and before the second troweling to allow the concrete to harden. When only a slight indentation can be made by pressing a hand against the surface, the second troweling should begin, using a smaller trowel with the blade tilted slightly.

If the desired finish is not obtained with the second troweling, a third troweling can be tried after another lapse of time. The final pass should make a ringing sound as the tilted blade moves over the hardening surface.

On large slabs where the finisher cannot reach the entire surface, he must work on knee boards. It is customary to float and immediately trowel (for the first troweling) an area before moving the kneeboards. This operation should be delayed until the concrete has hardened enough so that water and fine materials are not brought to the surface. Trowels must be kept clean and the edges protected from nicks when the tools are not in use.

Brooms

A smooth, troweled slab is easy to clean, but can be slippery when wet. For better footing, it can be roughened slightly by brooming to produce a nonslip surface.

The brushed surface is made by drawing a soft-bristled push broom with a long handle over the surface of the concrete slab after steel troweling. When coarser textures are desired, a stiffer-bristled broom can be used—for the stiffer textures this can be done floating rather than after troweling.

A broom drawn across the surface gives a coarse-textured finish. (Portland Cement Assn.)

A broomed texture can be applied in straight lines, curved lines, or wavy lines. Driveways and sidewalks are usually broomed at right angles to the direction of traffic.

Garden Water Hose

Curing of a concrete slab (or any concrete, for that matter) is one of the most important steps, but is often neglected. Concrete that has been properly mixed, placed, and finished will still result in a poor job if proper curing operations are not followed.

Concrete should be protected so that little or no moisture is lost during the early stages of hardening. Newly placed concrete should not be permitted to dry out too fast. And this is where the home garden hose becomes an invaluable tool: for moist curing of the concrete.

Water-curing with lawn sprinklers, nozzles, or soaker hoses must be continuous so that there is no chance of partial drying of the concrete during the curing stage.

Another common method is to cover the concrete with wet burlap. The burlap should be free of any substance that could harm the concrete or cause discoloration. The burlap can be placed as soon as the concrete is hard enough to withstand surface damage, and should be sprinkled periodically to keep the concrete surface moist at all times.

Curing should start as soon as it is possible to do so without damaging the surface. It should continue for 5 days in warm weather, or 7 days in cooler weather.

6. Planning Your Project

As with all home improvement projects, working with concrete requires advanced planning, preparation, and the proper tools.

Before you dig the first shovelful of dirt or mix the first batch of concrete, lay out the project on paper, establish dimensions, and estimate quantities of materials.

One of the first things to decide is whether you want to do it yourself or to hire professional help. If you take the job on as your own, you will have to determine whether you want to buy ready-mixed concrete or to mix the concrete on the job site.

Looking for a Professional

If you are building a new home, an ideal time to build and install concrete porches, sidewalks, driveways, and even patios is while the home builder has his work crews and subcontractors on the project. Most builders include walks and driveways in the basic home package. They usually subcontract the concrete work to specialists who have the equipment and experienced crews of cement finishers.

If you are adding improvements to an existing home—and especially if it is a fairly large project like a garage or driveway—it will pay to hire a concrete contractor.

Naturally, you want to select a reputable contractor and to assure yourself of a fair price for the job. It makes sense to get at least two different quotations on the job, making certain that both contractors bid on exactly the same specifications. The Better Business Bureau can tell you if a number of complaints have been filed against a contractor. If the contractor is a member of his trade association, this indicates his concern with the industry and many times means he has agreed to recognize industry standards and recommended practices. One good way to locate a contractor is to find a neighbor who is satisfied with the job done for him/her.

In selecting a contractor you should be concerned with more than just price. Talk to the contractor about concrete and ask his recommendations so that you will get some idea of his knowledge. If you are in a northern climate and he never heard of air entrainment, there is reason to question his knowledge of concrete. If he says joints are not needed, you also might wonder about his practical knowledge. It is a good practice to ask the contractor being considered about where he has built similar projects. Then visit these locations and examine the concrete job.

Design Plans

Building Permits

Almost all municipalities have zoning regulations and building codes, and require a building permit before remodeling or new construction begins. So before you begin construction, and when drawing up plans, check with your local building department. The building department usually inspects construction to make sure it conforms to the building code, and often permits are issued only after approval of the plans.

Layouts

Whether it is a utilitarian sidewalk or backyard garden, it pays to measure the area and then draw a layout plan.

For the yard, or outdoor living area, make a sketch on which you locate the house and garage, trees and shrubs within the lot lines, and other fixed elements. Make several prints (on a copying machine) of this basic sketch. Then make your rough sketches on these prints. Try several ideas on paper to see which looks the best. Then estimate costs. In the compromise between what you want and your budget, you may want to divide your overall plan into parts and set priorities on what is built first. And, of course, if you plan it as a do-it-yourself project, you will want to divide it into a series of small jobs and build it as your time and strength allow.

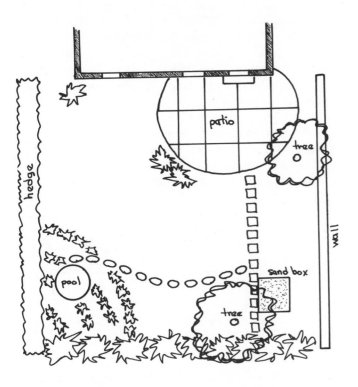

Lay the backyard out on paper, and select the combination of natural and concrete landscaping that suits you and your house.

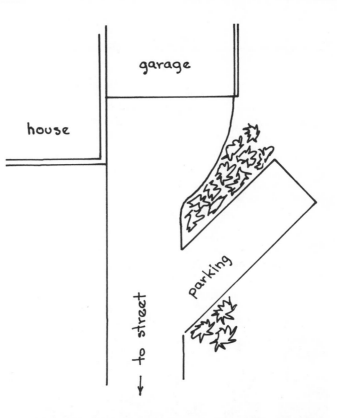

Top: A single-car width driveway should be widened near the garage to provide adequate access; this design also allows for off-street parking. Bottom: This driveway design allows space to turn a car around and easy access into the garage.

Driveway Plans

The plans for the driveway are dictated in large part by the location of house and garage on the lot. However, there are several other aspects to take into account. Consider off-street parking. For the multi-car family, and also for guests, extra parking space comes in handy. Good planning will enhance ease of entry from auto to house, and also offer plenty of room for maneuvering and for guest parking. Remember that a paved surface has a multitude of uses other than just parking space; the drive and parking area can also double as a play area.

A driveway should be as straight as circumstances permit, for curved drives are somewhat difficult to negotiate, particularly when backing up. If possible, when laying out the driveway allow enough space to turn your car around. This is especially desirable if your home is on a busy street, where backing out into heavy traffic can be dangerous.

Width. Single-car-width driveways are usually 10 to 14 feet wide. Curving drives need to be wider than straight drives, and should be at least 14 feet

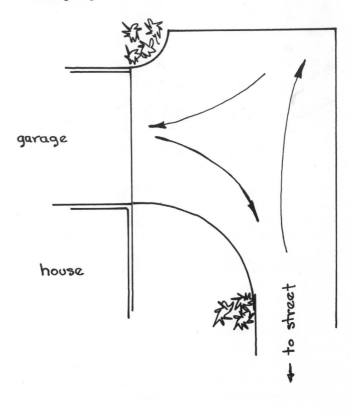

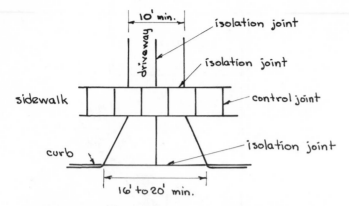

Plan for joints between drive and street and also where the driveway meets the sidewalk. A two-car width driveway should be built with a longitudinal joint down the middle.

This driveway provides both turn-around space and extra parking space. (Portland Cement Assn.)

Street Access. The driveway entrance must meet the local code as well as good commonsense safety criteria. Hedges and other foliage should not block the driver's vision when leaving the drive or entering the street. The driveway should meet the road or street at shoulder edge or gutter elevation; it must also meet the sidewalk grade if it crosses a sidewalk. The entrance should flare out for easy entry. Be sure to check with your local municipality before you build the driveway entrance or break out a curb; since the driveway enters a public thoroughfare it must meet local requirements.

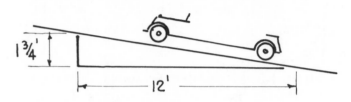

Maximum grade for a driveway should not exceed 14 percent—1¾ inches per lineal foot. In 12 feet the rise should not be more than 1¾ feet.

Grade and Thickness. Grade is another important factor, as is drainage. If the change in grade is too sudden the car's bumper or underside will scrape. If the grade is too steep, winter ice and snow can make the drive too slippery to negotiate. The grade should not exceed 14 or 15 percent (approximately 1¾-inch vertical rise for each horizontal foot of drive).

Colored concrete enhances the driveway and entrance as architectural features of the home. Note the gentle slope to the street so that the car does not scrape along the bottom. (Portland Cement Assn.)

wide. A driveway should be 3 feet wider than the widest vehicle it will carry. For a two-car garage, it is often impractical and uneconomical to build a drive that has the full two-car width, especially if it is a long distance from street to garage. Also, your lot may only have room for the single-car width. In such a case, carry the single-lane drive from the street to near the garage, where it must be widened for access to both sides of the garage. When the garage is at the back of the lot and the driveway runs close to the house, keep at least a 2½-foot clearance between the drive and the house. Short driveways for two-car garages can usually be built with full two-car width, about 16 to 24 feet wide.

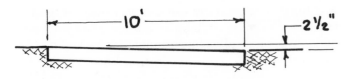

The driveway should drain away from the house. A slope of ¼ inch per running foot is sufficient; for a 10-foot wide driveway, the slope is 2½ inches.

The driveway should also be built with a slight slope so that it will drain quickly after a rain. The driveway should slope away from the house in the short dimension; in the long dimension the drive should slope away from the garage and toward the street. A slope of ¼ inch per running foot is recommended; that is, for a 10-foot wide drive, the slope would be 2½ inches.

If the driveway is expected to support only automobiles, a 4-inch thick slab is usually sufficient. If a fuel truck or other heavy loads are expected, it is better to make the slab 5 or 6 inches thick.

Sidewalk Plans and Layout

Residential sidewalks might be said to fall into four categories: public walks, front entrance walks, service walks, and garden paths.

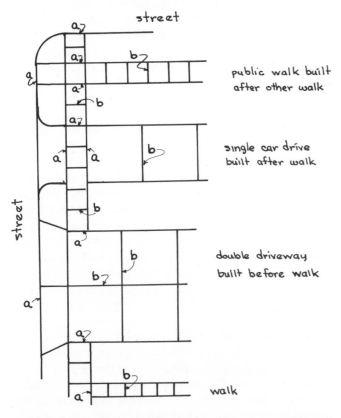

Isolation joints (a) should be located where a sidewalk and drive meet or at the junction of street or curb with a drive or walk. Control joints (b) are grooved into the slab to predetermine the location of cracks.

The public walk in residential areas is usually laid along the property line and the local municipality establishes the specifications. Depending on street right-of-way, the walk is ordinarily located 6 to 16 feet from the curb, leaving a strip of lawn between walk and curb. Public walks should be at least wide enough to allow two people walking abreast to pass a third person without crowding. The typical residential walk is 3 to 5 feet wide.

Front entrance walks are typically 3 to 4 feet wide. They can be straight or curved. Where trees are in the way, walks can be curved around them. When building near trees, allowance should be made for growth of the trees—otherwise the walk may be heaved and broken up by the spreading roots. For laying out pleasing curves in walks, a flexible garden hose can be used to mark off the curvature.

A slope of ¼ inch per foot will take care of surface drainage. Slope sidewalks away from buildings.

Walks are ordinarily 4 inches thick. Garden paths need not be as thick; 3 to 4 inches would be adequate, and many precast slabs come 2 inches thick.

Service walks connecting a back door and garage, or utility area, can be 2 to 3 feet wide.

Planning the Patio

Patio designs are limited only by your imagination, and budget. As mentioned earlier in this chapter, first lay it out on paper.

Consider the space available for your patio. This small patio features exposed aggregate concrete divided by redwood strips. (Portland Cement Assn.)

In designing the patio, consider the space available and how you plan to use your backyard—in most cases for family recreation and entertaining—keeping in mind privacy and landscaping. A patio tends to draw the family out of doors and functions as an outdoor living room, dining room, and play area.

Landscaping can be a part of a patio plan. You can use plantings around the edge to "soften" the patio, and planting pockets can also be built into the patio area. Trees and shrubs offer both beauty and privacy for family and guests.

A patio can be connected to the house or built as a special area in the lawn. Lot size and how the house is situated on the lot will affect location of the patio. Overall patio planning should include consideration of such factors as the view, climate, traffic flow to the house, weather, privacy, and how you plan to use the patio. Getting the best use of shade and sun can add hours of enjoyment. A patio on the south side of the house will get the sun all year round. Patios facing west are likely to be hot in the afternoon and cool in the morning. In hot climates, patios facing east are ideal because they cool off in the afternoon. North-facing patios never receive direct sun.

Plot the traffic flow to and from the house and patio when laying out possible locations. Positions of doors and windows in relation to the patio not only relate to traffic flow, but to the view possible and the convenience of bringing indoors outdoors, when entertaining and cooking.

Privacy needs will be determined by location on the lot, street traffic, and distance to the nearest neighbor. Privacy screens can be walls, fences, or other barriers such as shrubbery or vines.

A patio can be any shape: square and rectangular patios are common; curved and free-form patios are especially attractive. Any of these plans can be attractively accented with plantings of shrubs and flowers.

How Much Concrete Do You Need?

To estimate the amount of concrete needed, in cubic yards, you multiply the area (in square feet) by the slab thickness (in fractions of a foot) and divide by 27.

$$\frac{\text{Width (ft.) x Length (ft.) x Thickness (ft.)}}{27} = \text{Cu. Yds.}$$

For example, to find the amount of concrete needed for a 4-inch thick driveway that is 10 feet wide and 12 feet long, first multiply the length times the width (10 feet x 12 feet = 120 square feet). Then multiply by the thickness (4 inches equals ⅓ foot) and divide by 27.

$$\frac{120 \times ⅓}{27} = 1.48 \text{ cubic yards}$$

To allow for variations in the subgrade and for wastage, you should figure you will need 2 cubic yards. It is better to have a little too much concrete than not quite enough. If you are ordering ready-mixed concrete, many producers have a minimum amount they deliver or charge for—usually 2 or 3 cubic yards. In fact, most truck mixer drums require at least 2 cubic yards to achieve a thorough mixing.

How to Order Ready-Mixed Concrete

The ready-mixed concrete supplier can do a better job in supplying concrete if you give him complete information on your job in terms of specifications for the concrete.

The ready-mix supplier will need to know how many cubic yards of concrete are required, what kind of concrete you want (specifications), where to deliver it, and when. Place your order at least a day ahead of time, preferably several days ahead. Tell the supplier whether the truck can drive across the yard to discharge concrete directly into the forms or will have to park on the street to discharge into wheelbarrows. The dispatcher needs this information so he can schedule unloading time; you may be charged extra if you require excessive unloading time.

In specifying concrete, you may simply tell the ready-mixed supplier what it is for—a patio, for example—and rely on him to furnish the proper concrete. The knowledgeable supplier knows the climate and expected exposure, the normal loading expected for various structures and the properties of his ingredients, and so can tell you what concrete mix will work best for your job. However, it is preferable to indicate your general specifications: compressive strength, slump, maximum size of aggregate, and air content. You may also wish to specify cement content, although if you have specified the above four factors you will have given sufficient information for the supplier to furnish the proper mix.

By the way, the ready-mix supplier cannot guarantee strength of the concrete in the structure or slab since he has no control over all of the things on the construction site which can affect concrete.

But he knows which mix will meet your specification for strength if properly placed and cured, and will attempt to furnish that mix.

Specifications for Ready-Mixed Concrete

The fundamentals of mix proportioning (mix design) given in Chapter 4 are applicable to both ready-mixed concrete and do-it-yourself mixing at the job site. But for those who might prefer some standard recipe for usual home projects—patios, walks, driveways—here are two typical recipes.

Specifying Concrete Strength and Cement Content by Weight. As noted earlier the preferred method for proportioning mixes, and the most accurate, is by weight. For concrete with a compressive strength (at 28 days) of 3,500 pounds per square inch, the Portland Cement Association suggests the figures in the accompanying chart.

Guide for Ordering Ready-Mixed Concrete for Drives, Walks, and Patios*

Maximum size aggregate, inches	Minimum cement content, pounds per cubic yard	Maximum slump, inches	Compressive strength at 28 days, pounds per square inch	Air content, percent by volume
⅜	610	4	3500	7½ ± 1
½	590	4	3500	7½ ± 1
¾	540	4	3500	6 ± 1
1	520	4	3500	6 ± 1
1½	470	4	3500	5 ± 1

*Source: *Cement Mason's Guide to Building Concrete Walks, Drives, Patios, and Steps,* Portland Cement Association, 1971.

A four-inch slump gives a good workable mix, easily placed and finished by hand, but not too wet or soupy. Maximum size aggregate should not be larger than one-fourth to one-third the slab thickness...for most residential slabs (and steps, too) a one-inch top size is suggested. The air contents shown insure the durability needed when concrete is exposed to freezing ahd thawing and de-icing salts. Also, for concrete so exposed, it is advisable to use a minimum cement content of 560 pounds per cubic yard.

Specifying Concrete by Cement Content. At one time it was quite common to specify cement content by sacks of cement per cubic yard of concrete.

Ready-mix suppliers are accustomed to such specifications. Although somewhat over-simplified, here is a mix which is suitable for most home jobs and provides adequate strength. For concrete with a 4-inch slump and a maximum size of coarse aggregate of 1 inch, in mild climates call for a concrete with 5 sacks of cement per cubic yard. Where concrete is exposed to freezing and thawing or de-icing salts, specify 6 sacks of cement per cubic yard. As given in the preceding table, durability is enhanced with 6 percent entrained air. No more than 6 gallons of mixing water should be used per bag of cement.

Be Ready for the Concrete

If you are mixing concrete on the job site, you are in the ideal spot to assure that all preliminary work is completed before you start placing concrete. It is even more important when using ready-mixed concrete.

You will save time and money if you are ready for the ready-mix truck when it arrives. This means having all preliminary work done and sufficient tools and help on hand to handle the concrete.

Have the formwork and sub-base ready. The formwork must be adequately braced. For slabs the subgrade should be compacted and dampened. You will need help shoveling the concrete into place and finishing it before it sets up. One or two able-bodied friends will come in mighty handy.

How close to where the concrete is wanted can the ready-mix truck drive? Getting it as close as possible will save a lot of work.

Ready-mix trucks weigh from 18 to 22 tons. Protect your lawn (and be sure the lawn is dry), driveway, or sidewalk by laying down 2 x 12-inch planks for the truck to drive on. If the truck is to be driven between the driveway forms and over the subgrade, use planking to prevent deep ruts.

You will need wheelbarrows if the truck cannot drive right to where concrete will be placed. Deep-tray contractor's wheelbarrows can be rented. Be sure to use planks under the wheelbarrow; it will make your job of pushing it easier.

Do not wait until late in the day to start. And, once the concreting job has begun it usually cannot be interrupted.

Since you may have a little excess concrete, it would be a shame not to take advantage of the material you paid for. So, why not build some extra forms for such things as stepping stones, garbage can bases, etc., to use the extra concrete instead of wasting it?

7. Driveways

This driveway, with its exposed aggregate finish, blends into the natural landscaping. The design also allows for off-street parking and turn-around space. (Portland Cement Assn.)

Driveways can also be built of precast concrete slabs. The same rules for subbase, drainage, and slope apply to both precast and cast-in-place driveways. (Portland Cement Assn.)

No other construction material can offer the homeowner the durability, the good looks, and the long-term economy that a concrete driveway can offer. Good concrete driveways are not difficult to build, and it costs no more to build a good driveway than to build and maintain a poor one.

The following pointers will help you build your driveway, or guide you in selecting and working with a contractor.

Design

Review the points covered in Chapter 6 in designing your driveway. Remember that its slope (transverse as well as longitudinal) should be about $3/16$ to $1/4$ inch per running foot for good drainage. However, the grade should not exceed 14 percent ($1\frac{3}{4}$ inch vertical rise per running foot). To avoid scraping your car's bumpers or tailpipe, do not make abrupt changes in grade near the street. At the curb, the driveway should be level with the existing street.

Control joints can be spaced at intervals equal to the width of the driveway, but never more than 20 feet apart. A good general spacing is 10 to 12 feet. The control joint is a sawed or tooled groove in the slab to regulate the location of cracks. A tooled joint should be cut $1/5$ the thickness of the slab...for a typical driveway slab, about $\frac{3}{4}$ inch to 1 inch. Isolation joints should be provided wherever the driveway borders a street, garage, sidewalk, or building. Isolation joints (also called "expansion joints") separate the drive from other structures to accommodate horizontal and vertical movement. Isolation joints should be about $\frac{3}{4}$ inch wide and extend the full depth of the slab. They are filled with compressible material, usually premolded filler. The premolded material should extend the full depth of the slab and not protrude above it.

The thickness of the slab will depend on the subgrade upon which it is to be cast, and the expected loads; however, a driveway should never be less than 4 inches thick. If trucks will drive on it or if located on poor soils, the driveway should be

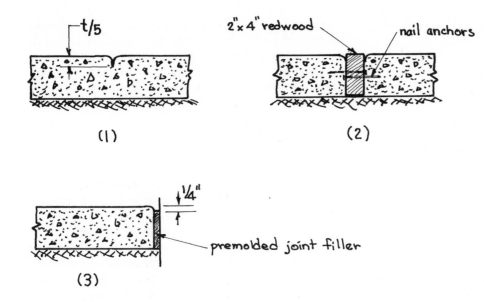

(1)

(2)

(3)

Types of joints used in concrete driveways, sidewalks, and patios: (1) Control joint—The tooled joint should be cut one-fifth the thickness of the slab. It forms a plane of weakness which predetermines the location of cracks. (2) Wood strips— These may function as decorative dividers as well as joints. The wood divider is anchored to the concrete by nails driven through the wood. (3) Isolation joint—This joint separates the slab from a structure or other slab. A premolded joint filler extends the full depth of the slab.

5 inches thick or more. Steel mesh reinforcement, properly positioned, will help to minimize the width of potential cracks.

Materials Needed

Concrete

Fundamentals of proportioning and making concrete are covered in Chapters 3 and 4. As a general rule, to achieve a durable driveway, specify concrete as follows:

Strength: 3000 to 3500 pounds per square inch;
Slump: 4 inches;
Coarse aggregate (maximum size): ¾ to 1½ inch depending on slab thickness;
Air content (if exposed to freezing and thawing): 6 percent plus or minus 1 percent.

Subbase

Proper preparation of the subbase is one of the keys to successful driveway construction. It is the subgrade which supplies structural strength to the driveway slab. The slab itself cannot bridge portions of the subgrade that settle or are washed out, and still carry the weight of an automobile without cracking.

The subbase must be well drained. It should also be level, hard, and free from foreign matter. Soil should be removed to the depth required for the slab. Soft sod, vegetable matter, and debris should be cleared from the ground. Soft spots should be dug out and filled with at least 2 inches of sand, gravel, or crushed stone, and thoroughly tamped down.

If the soil is firm and well drained it is possible to build a slab right on it. In fact, undisturbed soil is superior as support for a concrete slab to soil that has been dug out and the fill poorly compacted. However, a good base can be created by digging out the sod and soil deep enough for a slab, and filling with an adequate granular fill of sand, gravel, slag, or crushed stone. In building up the fill, use a sound granular material and compact it thoroughly in layers of no more than 4 inches in thickness. Subgrade compaction can be done with hand tampers or rollers.

The bottom of an undrained granular base course should not be lower than the adjacent finished grade; otherwise the base course becomes a reservoir for water.

Forms

Usually 2x4-, 2x6-, or 2x8-inch lumber set on edge is used for driveway formwork, depending on slab depth desired. The side forms should be rigidly braced with stakes on the outside of the forms. The stakes can be either 2x2 or 1x2 inches, 18 inches long, and placed about every 3 running feet of board length. Drive them firmly and deeply into the ground and butt them up right next to the forms. Nail the stakes securely to the side forms. Steel stakes are also available, and are easy to use and have greater holding power. Tamp earth along the outside of the form to help keep the boards secure during placing and finishing operations.

*stake removed
after nailing to outside stakes*

Curved forms can be built with strips of sheet metal, hardboard, or plywood. Plywood ¼ to ½ inch thick should be bent with the grain vertical. The stakes need to be set closer together on curves than on straight forms, at about 1- to 2-foot intervals. The wood can be bent by setting outside and inside stakes. Nail the form to the outside stakes, and the inside stakes can then be removed.

Forms and screeds should be set flush with the soil level—or to a predetermined grade—and set to true line. Drainage should always be provided; to prevent ponding on the surface of the drive the minimum grade should be 1 inch per 4 feet. Slope the drive away from the garage and house. Most of the layout can be handled with measuring tape or carpenter's rule, string, and a string level. For complicated layouts or grades you may want to use a builder's level.

Stakes will be required at each joint in the side form. A 1x4-inch stake should be used to lap joints; a 2x4-inch stake to maintain alignment of the ends. For curves, use ¼-inch plywood or sheet metal that has been staked firmly in place; this requires more stakes per running foot than the straight forms. Plywood for curves should be cut with the exterior grain vertical once the forms are in place.

Position any type of lumber to form isolation joints and fasten it securely (tie down or brace or nail). This board is later removed and a filler material inserted. Often, it is more convenient for the isolation joint to secure a premolded filler in place before the concreting, and then leave it.

Reinforcing Wire or Mesh

The value of reinforcement in driveway construction has been, to a degree, open to question. Some consider the cross-sectional area of steel insufficient to add to the strength of the slab, and say that more good comes from adding another inch of plain concrete to slab thickness. Those in favor of wire mesh reinforcement insist the mesh minimizes cracking and crack sizes, and helps hold the slab together. The author's preference is for welded wire fabric or mesh in the slab.

The most common size of welded wire fabric used for driveway construction is 6x6—W2.9x W2.9* (formerly designated 6x6-6/6), stocked by most building supply dealers in 5- or 6-foot-wide rolls or sheets.

Reinforcement for a driveway slab should be located at or slightly above the center of the slab (about 1½ inches from the top surface is a good location). If sheets of fabric are used, they should be overlapped by at least one wire spacing to achieve continuity. (Note: In structural slabs where steel is designed and located to add to structural strength, the reinforcement is usually placed near the bottom. However, in a driveway slab the reinforcement is

An easy way to reinforce a concrete slab is to place mats of welded steel wire fabric. Here the bottom layer of concrete was placed and the reinforcing laid on that, then the remainder of the concrete placed and screeded.

A bar chair supports the reinforcing mesh in the slab where it belongs, rather than on the bottom against the subgrade. This plastic chair holds the mesh securely and the round base ring keeps it from sinking into the subbase. (Lotel Incorporated)

---

*The first figure (6x6) indicates spacing between wires in inches and the second (W2.9) indicates wire gauge.

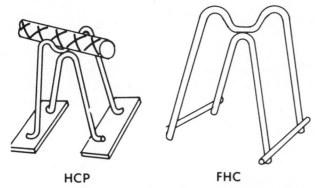

HCP   FHC

There are a wide variety of wire bar supports (chairs) to hold reinforcement. For those used in slabs on ground, a plate or wire is attached to the bottom to keep the support from sinking into the soil or subbase. (Reproduced from Manual of Standard Practice for Detailing Reinforced Concrete Structures (ACI 315-74), *American Concrete Institute, 1974)*

PLAIN PRECAST CONCRETE BLOCKS

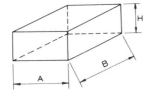

A	B	H
2″	2″	¾″
2″	2″	1″
2″	2″	1-½″
2″	2″	2″
3″	3″	2″
3″	3″	3″
3″	3″	4″
4″	4″	3″
4″	4″	4″
6″	6″	3″

PRECAST CONCRETE BLOCKS WITH WIRES

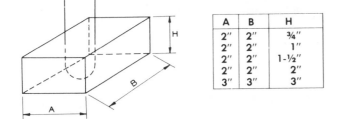

A	B	H
2″	2″	¾″
2″	2″	1″
2″	2″	1-½″
2″	2″	2″
3″	3″	3″

PRECAST CONCRETE DOWELED BLOCKS

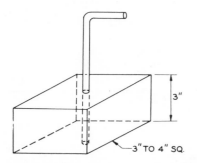

3″

3″ TO 4″ SQ.

Precast concrete bar supports are available in three styles: (top) plain; (center) with wires; (bottom) doweled. (Reproduced from Manual of Standard Practice for Detailing Reinforced Concrete Structures (ACI 315-74), *American Concrete Institute, 1974)*

used mainly to counteract the effects of thermal variations, which are pronounced near the top, so the steel is best placed near the top of the slab.)

To keep the reinforcing mesh in the top half of the slab, it should be supported on "chairs." These are made of steel or plastic. Makeshift ones are sometimes made from bits of concrete block, brick, or stones. Or a layer of concrete can be placed and the wire laid on that before concreting the top half. This latter method calls for quick action so that the layers of concrete bond together and the concrete does not set up before it can be placed and finished. Placing reinforcing mesh on the subgrade, and prying it up as concrete is placed, will not locate the steel in the slab where it will function as intended. If you plan to do this, it is better to leave out the steel and add another inch of concrete to the slab thickness, and get more for your money and effort.

Preparing for Concreting

Before concreting begins, check all forms for trueness to grade and proper slope for drainage, and also to see that forms are securely anchored.

Make sure that the subgrade is smooth and well compacted. You also want to ensure a correct, even slab thickness. This can be checked with a string line and rule. Or, prepare a template of lumber equal to the desired slab thickness. Holding the top level with the top of the form, determine any deviations in the subgrade. These should be smoothed out or filled in for uniform slab thickness. The subgrade should be thoroughly dampened with water in advance of concreting; but, there should be no free-standing water or muddy or soft spots at the time the concrete is placed.

To make it easier to remove forms after concreting, dampen the forms with water, or oil the forms. Motor oil can be used. You can also buy a regular form-release agent.

If using reinforcement, have it in place on bar chairs before the ready-mixed concrete truck arrives.

Placing Concrete

If possible, the truck should discharge the concrete directly in place. Usually this is not feasible for residential work, and concrete can be transported in wheelbarrows. So make sure your helpers are on hand and ready.

Move the concrete horizontally as little as possible. Spread it evenly with a short-handled, square-end shovel to full depth in the forms, and approximately 1 inch above the forms to allow for compaction. Thoroughly consolidate the concrete by using spades around the perimeter of the slab next to the forms, and hand tampers over the remainder of the slab-work. Two or three passes with a hand tamper will usually compact the concrete to a point virtually level with the top of the forms.

Strike off the excess concrete by screeding in a saw-like motion with the straightedge resting on the side forms. However, do not overwork the concrete. As the straightedge is pulled over the surface, the low spots left behind it must be filled by placing additional concrete in them with a shovel, carefully avoiding separation. These spots are then rescreeded.

Darbying should immediately follow screeding. It should be completed before any excess moisture or bleeding water is present on the surface.

Any finishing operation performed while there is excess moisture or bleeding water on the surface will almost certainly cause dusting or scaling, and your driveway surface will deteriorate. Darbying prepares the surface for subsequent edging, jointing, floating, and troweling (all discussed in Chapter 5).

Now you must wait...a slight stiffening of the concrete is necessary before proceeding further.

Finishing

Concrete should be finished as little as possible—just enough to smooth out any irregularities and to compact the surface. Wait until the concrete has set up sufficiently to support a man without leaving more than about ¼-inch deep impressions of his heels.

After all bleed water has disappeared from the surface and the concrete has started to stiffen, edging can be started.

Immediately following edging, or at the same time, the slab is jointed or grooved. Cut all control joints to a depth of one-fifth to one-quarter the thickness of the slab—in general, ¾ to 1 inch deep. If the slab is to be grooved only for decorative purposes, jointers (groovers) having shallower bits may be used.

If the driveway is 20 feet or more in width, place an expansion joint along the entire length in the middle. To do this efficiently, place one half of the driveway at a time, using a bulkhead for the center joint. After the second half is placed, replace the bulkhead with a premolded bituminous filler.

After edging and jointing operations, the slab should be floated. Many variables—concrete temperature, air temperature, relative humidity, and wind—make it difficult to set a definite time to begin floating. However, when the water sheen disappears and the concrete will support a man, it is ready to be floated.

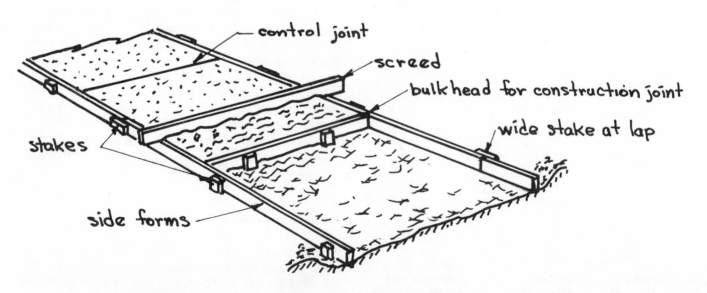

The screed or straightedge is worked back and forth to even off the concrete. The side forms should be staked securely. Use a wide stake where pieces of side forms are joined. When construction stops for the day, use a bulkhead to hold concrete; this can also function as a construction joint.

Procedures in Pouring-Finishing Slabs are given in this series of a dozen photos. In the first group, the procedure begins after forming and grade preparation. First the fill-in gravel base is tamped to compact it (1) followed by wetting down the grade (2). Wet concrete from the ready-mix truck chute is then dumped within the forms and spread using a square shovel (3). Leveling is followed by see-sawing a strike-off board to bring the surface smoothly even with the top edges of the form boards. Then, the finisher uses a large bullfloat (4) to fill small voids and raise the concrete fines to the surface.*

In the second group of photos, concrete finishing is done with a set of specialized finishing tools used in the following order: hand float (5) to bring the fines further to a smooth surface covering the larger aggregate pieces; then a mason's pointed trowel (6) is used to slice along the form's side breaking the concrete contact for the top couple inches thus preparing for the run along the form

with the edger (7) which gives the concrete a smoothly rounded top corner. At regular intervals along the walk or slab, usually from 3½ to about 6 feet, a joint is cut using a board for a straight-edge guide (8). The jointer tool is designed to place a slightly rounded V-groove in the concrete which is about ¾ inch deep.

Final finishing work is indicated in the final group of photos. After edging and jointing comes the working of the slab's surface which is by now drying out and becoming free of surface water. A float (9) first roughs up the setting surface slightly giving it a properly moist and textured top ⅛-inch which is then brought to a final smooth and slick surface by a steel finishing trowel (10). The finisher works the surface using the trowel at a slight tilt and with even pressure in a sweeping motion. Back and forth across in a semi-circle smoothing out previous trowel marks. Final troweling is best accomplished when the top surface looks nearly dry. Outdoor walks and drives left trowel smooth are apt to be slippery underfoot in wet or snowy weather. Broom trowel work (11) for a slightly coarse texture. Last step (12) uses strips of burlap kept wetted down to cure the slab for a period of 2 or 3 days.

The marks left by the edger and jointer will be removed by floating. If these marks are desired for decorative purposes, the edger or jointer has to be rerun after the floating operation.

Immediately after floating comes troweling. The purpose of troweling is to produce a smooth, hard surface (which also becomes quite slippery when wet). Sometimes driveways are troweled, but it is seldom necessary to steel-trowel finish an outdoor driveway. A coarse, nonslip surface can be produced with a wood float. A rougher, textured surface can be produced by brushing or brooming transversely across the slab: the brushed surface is made by drawing a soft-bristled push broom over the surface; when coarser textures are wanted, a stiffer bristled broom may be used.

Curing

Curing may be the most critical step in driveway construction, since it is the key to durability. Curing often makes the difference between a highly satisfactory job and one that cracks up and performs poorly. Lack of moist curing can cut the potential strength of concrete by as much as one-half.

The process is simple; all you need to do is: (1) keep the concrete surface moist by spraying, moistened burlap, etc., or (2) prevent the loss of moisture from the slab by laying curing paper, or polyethylene film, or by spraying a membrane curing compound on the surface.

Curing should be started as soon as it is possible to apply the curing medium without damaging the surface. If concrete surface temperatures are 70 degrees F. or higher, curing should continue for at least 5 days. At temperatures of 50 to 70 degrees F., curing should continue for at least 7 days.

Curing with water is the most effective way. Burlap is the most commonly used wet covering. It should be laid over the concrete as soon as the concrete is sufficiently hard to withstand surface damage. Cover all exposed concrete edges. Keep the covering continuously moist throughout the curing period.

If sprinkling is employed, use a fine spray applied continuously rather than intermittently, so that there is no chance of partial drying of the surface (which can result in crazing and cracking). Sprinklers or soaking hoses can also be used.

The other curing method is moisture retention; that is, utilizing the moisture within the slab by sealing the surface. Sprayed-on curing compounds can be applied with a hand sprayer. The membrane should be applied when the surface of the concrete is still damp but not wet with free water. Polyethylene film and waterproof paper are other ways of covering a slab to retain moisture. Plastic membranes and waterproof paper should normally be avoided for curing colored slabs, because they cause uneven moisture distribution over the top surface of the concrete and result in a blotchy surface.

Decorative Effects

It is easy to build a driveway, and just as easy to add decorative effects at very little additional cost. A great number of designs and color effects are possible.

Examples of decorative effects would include some of the following (some simple, others complex).

Brooming. Instead of the usual straight, transverse strokes, a zig-zag pattern can be used. Or squares can be set off by brooming in opposite directions.

Designs. Attractive geometric designs can be stamped into the surface.

Color. Color can be incorporated in the concrete. If alternate slabs are cast, contrasting colors can be used.

Aggregates. An exposed aggregate can be rolled into all or part of the surface of the partially hardened slab; or the surface can be brushed before it is hardened to expose the coarse aggregates.

See Chapter 11 for details on color and other decorative treatments.

Checklist for Driveways

1. Prepare the subgrade thoroughly.
2. Be sure that good drainage is provided.
3. Use an air-entrained concrete mix if concrete is exposed to freezing and thawing.
4. Provide adequate, properly spaced joints.
5. Finish with a skid-resistant texture.
6. Cure for at least 5 days.

8. Sidewalks

The construction of a high-quality, durable concrete walk is surprisingly simple. You can mix concrete on the site and place sections of the walk in easy stages at your convenience. If ready-mixed concrete is ordered, you will want to complete the job at one time; this may mean getting extra help so that all the concrete can be properly placed and finished. Or you can use precast slabs of a wide variety of shapes and textures.

The main walk should be at least 3 feet wide, and service walks a minimum of 2 feet wide. In most cases, a sidewalk slab thickness of 4 inches is adequate.

For a 4-inch-thick sidewalk, 1 cubic yard of concrete will be enough for 27 lineal feet of widewalk, 3 feet wide.

A good rule-of-thumb is that control joints should be cut at intervals equal to the width of the sidewalk slab, and never more than 10 feet apart.

In general, sidewalks are constructed in the same manner, and with the same concrete mixes, as driveways.

An attractive exposed aggregate sidewalk with wood divider strips leads up to this home. This house also features precast concrete roof channels and concrete split block walls. (Portland Cement Assn.)

A wide front entrance walkway features exposed aggregate concrete, along with wide divider strips of colored concrete and narrow dividers of redwood. (Portland Cement Assn.)

Precast colored concrete brick are laid between cast-in-place concrete borders. (Portland Cement Assn.)

Materials Needed

Concrete

Refer to Chapters 3 and 4 for content and preparation of mix.

A typical sidewalk mix, proportioned by weight, would consist of 94 pounds of cement, 215 pounds of sand, and 295 pounds of coarse aggregate (gravel or crushed stone). Or, if proportioned by volume, the mix is composed of 1 part cement to 2¼ parts sand and 3 parts coarse aggregate.

Use only enough water to produce a concrete that works easily and does not separate. Under ordinary conditions, use a water-cement ratio (by weight) of around 0.50; that is, one pound of water for every two pounds of cement; or 5½ gallons of water per sack of cement.

Air entrainment is essential in northern climates. It pays off in strength and durability. Air entrainment is a primary deterrent to spalling and scaling especially where de-icers are used to melt snow and ice on walks. For a sidewalk mix with 1-inch maximum size aggregate, 6 percent air (plus or minus 1 %) is recommended.

Formwork and Layout

In laying out a sidewalk make certain it drains away from the house. A cross-slope of ⅛ inch per foot is adequate to assure drainage.

For laying out pleasing curves in a walk, a flexible garden hose can be used to mark off the curvature, and the sod excavated along the curve.

Curved forms are easily made with two thin strips of plywood or hardboard, as shown in the illustration. Bend the first strip to the proper radius by setting outside and inside stakes. Nail through the strips to the outside stakes. Then place the second strip inside the first and nail the two together. Remove the inside stake and the form will hold its curved shape.

For a 4-inch slab, the handiest way to build the side forms is to use 2x4-inch boards. These are held securely in place by stakes. Dampen or oil the forms

A curved walkway featuring a border of colored concrete. (American Concrete Institute)

prior to concreting; it makes it easier to remove forms after concreting.

Construction

For cast-in-place sidewalks you can follow the same construction procedures as given for driveways in Chapter 7. Just by way of review, here are a few basic steps.

All sod and debris must be removed to the depth of the walk. A subbase of crushed stone or other granular material is used where the soil is spongy or the area low and wet. Tamp the subbase thoroughly to compact the fill.

Dampen the subgrade, but not so much as to leave standing water.

Place the concrete in the forms. To assure a compact surface, leave the newly placed concrete slightly higher than the intended finish level. Once it is placed, the concrete is tamped. Tamp it thoroughly. Spade along the edges to eliminate air pockets.

Level the tamped surface with a strikeboard or straightedge. If necessary, fill any depressions with extra concrete. Round off the sides with an edger to prevent chipping.

Control and isolation (expansion) joints are placed to prevent unsightly and irregular cracks. Isolation joints should be placed wherever the sidewalk meets another walk, driveway, curb, building, or wall. The joint consists of premolded material about ½ inch thick and extends the full depth of the walk. Control joints, which are weakened planes to con-

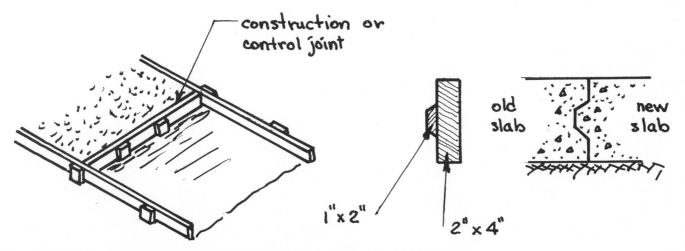

When forming a construction joint with a keyed bulkhead, use the keyed control joint to aid in keeping the two slabs level with each other.

trol cracking, are cut with a groover 4 to 5 feet apart to a depth of $1/5$ to $1/4$ the slab thickness.

To insure safety in wet weather, sidewalks should be finished with a slightly rough, gritty surface for a nonslip surface. Finish off with a wood float. You can also draw a hair broom across the concrete before it has set and achieve about the same kind of finish.

Pattern the walk, if desired, before the concrete hardens. Decorative effects are covered in Chapter 11.

If you do the walk in sections, place a wood bulkhead where concrete work stopped. The bulkhead is made with a 2x4-inch piece of wood cut to the inside dimension of the walk forms. A beveled 1x2-inch strip nailed the length of the 2x4 will form a keyed control joint so that future sidewalk slabs will remain level with the previously cast slabs.

As with all concrete, strength and durability come with adequate curing—keeping moisture in the concrete during the setting and hardening process. Be careful, and do not let that sidewalk slab dry out rapidly! The same curing procedures listed for driveways can be used for sidewalks. Ideally, five to seven days of damp curing is the most desirable period, and at a minimum some kind of curing should be provided for at least three days, even if it is nothing more than sprinkling the slab at regular intervals.

Checklist for Sidewalks

1. Prepare subgrade thoroughly.
2. Use air-entrained concrete in northern climate.
3. Provide adequate, properly spaced joints.
4. Finish with a safe, textured surface.
5. Provide for proper curing.

9. Patios

A carefully planned and well-built patio is a valuable extension to the home and yard. Planning of details to keep in mind were discussed in Chapter 6. Patios can be built in place, like driveways, or assembled from precast units.

Cast-in-Place Patios

Many of the basic techniques and procedures used to build concrete driveways are used to build cast-in-place patios. After all, the construction of flatwork is essentially the same regardless of how that slab will be used.

Following your drawn-up plan, mark off the area. Remove sod and soil to the depth desired. The subgrade should be uniform, hard, free from foreign matter, and well drained. Serious cracks and slab settlement can often be traced to a poorly compacted subgrade. Dig out soft or mucky spots and fill with granular material. Tamp the fill thoroughly.

This deluxe patio, which fits into the natural beauty of the backyard, was built with exposed aggregate concrete and redwood strips. Screen block and a brick barbecue add useful and finishing touches. (Portland Cement Assn.)

A cast-in-place patio using concrete brick set at an angle for the flower bed border. (Portland Cement Assn.)

Simple, but different. Round concrete slabs lead to a low terrace with a larger patio slab. (Portland Cement Assn.)

Granular fills of sand, gravel, crushed stone, or slag are recommended for bringing the slab area to uniform bearing and final grade. Compact the fill in layers not more than 4 inches thick. It is best to extend the fill at least 1 foot beyond the slab edge to prevent undercutting during rains. For best subgrade compaction, dampen the sand or gravel fill.

Cast-in-place patios should be a minimum of four inches thick and built with a ¼-inch per foot slope for drainage away from the house.

Most patios are formed with 2x4-inch wood forms, although metal forms or 1-inch lumber can also be used. Stake and brace the forms firmly to keep them in horizontal and vertical alignment, and to hold up against the pressure of the fresh concrete.

For a four-inch thick slab, 1x4- or 2x4-inch lumber may be used. Wood stakes can be cut from 1x2-, 1x4-, 2x2-, or 2x4-inch lumber. Space stakes at 4-foot intervals for 2-inch thick formwork and at 2- or 3-foot intervals for 1-inch thick lumber. Re-usable steel stakes are also sometimes used.

For ease in placing and finishing concrete, drive all stakes slightly below the top of the forms. Wood stakes can be sawed off flush. Drive stakes straight and true to keep forms plumb. For easy stripping later, use double-headed nails through the stake into the form.

For curved borders, drive stakes on desired arcs and place thin strips of wood around the stakes. The arc is held in place by outside stakes.

Wood side forms and divider strips may be left in place permanently for decorative purposes. (These divider strips also serve as control joints, and thus minimize cracking.) For dividers use 1x4- or 2x4-inch redwood or treated lumber. You will want to protect dividers by covering the top edge with masking tape before concrete is placed.

Since the divider strips and outside forms will be a decorative part of the patio, take greater care than usual in assembling the forms. Miter corner joints neatly. Join intersecting strips with neat butt joints. Do not drive nails through the top of the permanent strips. However, the permanent strips should be anchored to the concrete. Anchor outside forms to the concrete with 16-penny galvanized nails driven at about 16-inch intervals through the forms at midheight. Do the same thing with interior divider strips, but drive the nails from alternate sides of the board. Drive nail heads flush with the forms.

Before concreting, check all forms for trueness to grade and to make certain the forms are anchored securely. To remove the forms easily after the concrete sets, dampen the forms with water or oil them

When building a patio with wood divider strips that will be left in permanently, cover the top surfaces with masking tape before concreting. This protects the wood from abrasion and staining while the concrete is being placed. (Portland Cement Assn.)

before concreting. Motor oil can be used, or you can purchase a form-release agent.

The most convenient and economical source of concrete is a ready-mixed concrete producer. If you mix it yourself, refer to earlier chapters in the book about materials and proportioning. One cubic yard of concrete will place 81 square feet of 4-inch thick patio.

Before delivery of the concrete, make sure you have the tools needed, and the manpower, to place and finish the concrete (see Chapter 5 on tools). If you have never finished concrete, consider hiring a concrete finisher. Plan to get the ready-mixed concrete truck as near as possible to the point of placement without driving over sidewalks or driveways. More often than not, you will find that you have to wheelbarrow the concrete in from the street.

Place concrete uniformly to the full depth of the forms, as near as possible to the final position. Start placing in a corner. Spade the concrete well into the corners and along the forms. Spreading and spading are best done with a short-handled, square-end shovel. Overfill the forms slightly, then screed with a straightedge. Screeding merely levels the concrete with the top of the forms. Screeding or strike off is followed immediately by darbying. Darbying levels ridges and fills voids left by the straightedge, and helps to embed coarse aggregate slightly below the surface.

If the forms are not to be left in place, immediately after using the darby, cut the concrete away from the forms to a depth of about 1 inch, using a pointed mason trowel.

Check the concrete frequently to determine when it is ready for finishing. Finishing operations include edging, jointing, floating, troweling, and brooming. Finishing operations must wait until all bleed water has left the surface (that is, when the water sheen is gone), and the concrete stiffens slightly (that is when the concrete can bear foot pressure with only about ¼-inch indentation).

Edging is the first operation in finishing. Delay preliminary edging until the concrete has set sufficiently to hold the shape of the edger tool. Edging produces a neat, rounded edge that prevents chipping when the form is removed. Edging also compacts the concrete surface next to the form. Edging is sometimes required after each finishing operation.

Concurrently with edging, or right after, the slab is jointed or grooved. Proper jointing can eliminate unsightly random cracks. For patios, the spacing of control joints should not exceed 10 feet in either direction. The panels should be approximately square. The smaller the panel, the less chance there is for random cracking. Control joints sould be continuous, not staggered or offset. The tooled control joint should extend into the slab one-fourth to one-fifth of slab thickness. This tooled joint provides a weakened section that induces cracking beneath the joint, where it will not be visible. And, of course, if wood divider strips are left in place for decorating purposes, these also serve as control joints.

The other type of joint is planned and built in prior to concreting—the isolation or expansion joint. It consists of a premolded strip of fiber material that extends the full depth of the slab or slightly below it. An isolation joint belongs where the patio borders the house. Isolation joints should be flush with the concrete surface, or about ¼ inch below.

The next step is floating. Floating produces a relatively even texture which has good skid resistance and thus makes a good final finish.

If a hard, dense surface is desired, the surface can be troweled right after floating. Troweling should be delayed until the concrete has hardened enough so that water and fine material are not brought to the surface.

Curing follows after finishing. Curing can be accomplished by covering with a plastic sheet, spraying on a curing membrane, covering with wet burlap, or soaking with water. Moist-cure the concrete for at least five days.

Precast Patio Slabs

Precast concrete patio slabs are readily available from local precast or concrete masonry producers and building materials centers. These units come in a wide range of sizes, colors, textures, and patterns. The units are easy to handle and install, requiring only a few tools and little special knowledge of concrete or finishing techniques. Another advantage is that the work can be spaced out to fit your schedule, and sections of the patio can be installed in small parts rather than built all at once.

Integrally colored square and rectangular patio blocks combine for an attractive patio that is also easy to install. (National Concrete Masonry Assn.)

There is one thing you can do with precast patios that you cannot do with cast-in-place—and that is redecorate. It is quite common to redecorate the living room, but now you can redecorate your outdoor living area with precast patio units. One Michigan specialist who builds patios reports that a few homeowners change the patio scheme every

Interlocking precast paving block offers special advantages for this pool patio in Canada. At the expense of one or two blocks, the slab can easily be taken up and relaid without spoiling its appearance if it becomes necessary to repair the underground services that are a part of a swimming pool installation. (KNR Concrete Systems Ltd. and North American Stone Co.)

several years by changing colors and patterns of their precast patio units.

Precast slabs are commonly available in square and rectangular shapes; other shapes are hexagonal, triangular, and round. Slab thicknesses vary from 2 to 3 inches. The 2-inch thickness is quite suitable for patios. Precast slabs are available in the normal gray cement colors, white cement color, and tones of red, green, brown. Exposed aggregate finishes are also available.

Installation

While precast slabs can be laid directly on ground which is firm and solid, it is preferable to lay the patio on a smooth sand bed.

Prepare the site. Remove sod, soil, and debris to a depth about twice the thickness of the precast slab. Where there are soft, wet areas, remove the soil and replace with 4 to 6 inches of granular fill of gravel, crushed stone, or slag. Tamp the fill material in place.

Sand bed. After cleaning the patio area, place a 2-inch sand bed. Dampen the sand and tamp it thoroughly in place. Level the sand bed using a 2x4-inch straightedge. You may wish to embed two straight boards in the sand to serve as screed guides and establish a level, smooth bed. Level the sand by running the straightedge over the leveling boards. After leveling and smoothing the sand, the boards can be removed.

Square precast patio slabs can be bought at garden centers, building supply centers, and concrete products plants. Installation is easy. Top: Stake off the patio area and string a line between stakes. Set dimensions in even modules to fit the size of the slabs so that cutting of slabs is not required. Excavate sod and soil to a depth equal to that of the slab thickness plus 2 or 3 inches for a sand subbase. Center: Spread a sand layer 2 or 3 inches thick and compact it thoroughly. Set screed boards in the sand to establish a level, smooth base. Also set a slight slope so that the patio drains away from the house. Bottom: Level the sand by running a straightedge over the leveling boards. After smoothing and leveling the sand, remove the screed boards. Opposite page: Set the precase slabs on the sand subbase by starting in a corner. Butt the joints together. After all slabs are placed, sweep sand into the joints to complete the job. (Portland Cement Assn.)

Laying the precast slab. Carefully lay the precast units to the pattern desired with the shapes selected. It is best to start in a corner. Level and make adjustments as necessary. Complete one row at a time.

The slabs can be set with joints butted, or leave ¼- to ½-inch space between units and sweep sand into the joints to complete the job. You can also fill the joints with an asphaltic mixture. (For a less formal, garden-type appearance, leave joints ¾ to 1 inch wide between slabs, and plant grass or ground cover in the spaces between.)

When all slabs have been put down, water the patio heavily; this will draw the slabs down tightly. Stay off the patio and do not walk across it until the patio has dried thoroughly.

Mold Your Own

You may wish to mold your own patio paving slabs. Such a job can be done at your leisure, and the finished precast units stacked until ready for installation.

You can build your own forms to whatever shape pleases you, or buy them ready-made. Several manufacturers sell forms primarily to precast concrete plants for mass production, but some of these forms might be adapted to do-it-yourself projects.

At least one manufacturer markets a mold specifically for the homeowner to mold his own slabs. The unit consists of a fiberglass mold and an ejector plate. It makes a hexagonal slab measuring 12 inches wide and 2 inches deep; the cast slab weighs approximately 16½ pounds. Here is how to make it.

1. Place the mold on a flat surface. Insert the ejector plate in the bottom of the mold. Fill the mold with concrete. (This manufacturer suggests a mix with 3 parts sand and 1 part cement, and just enough water so that freshly molded block will hold together without slumping.)

2. Tamp the concrete thoroughly and firmly to drive out trapped air and consolidate the mix. Level off the concrete.

3. Turn the mold over and eject the newly cast slab onto a sheet of polyethylene or several layers of old newspapers.

4. Clean the mold and ejector plate and repeat the process.

5. Keep the slabs slightly moist for a minimum of 3 days to assure adequate curing.

Paving stone mold forms a hexagonal slab. The mold (foreground) is composed of two parts: side and bottom form, and ejector plate. (Decor Molds)

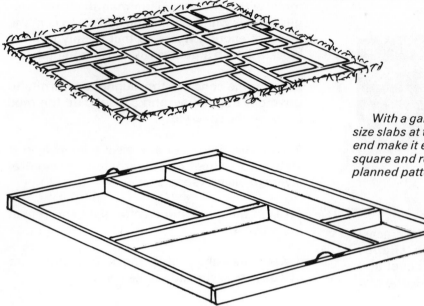

With a gang form, it is possible to cast several different size slabs at the same time. Screen door handles on each end make it easier to lift the form off after casting. The square and rectangular slabs can then be combined in a planned pattern. (top)

If fractional blocks are required, they must be cut to size immediately after demolding and before the concrete sets.

If you decide to build your own form, you can make it almost any shape. The simplest form is a square mold made up of 1x2-inch boards. The boards are cut to size and nailed together to form a square bottomless box. The box should be given a slight taper toward the top so that it can easily be lifted off the finished casting. For example, if the bottom of the box is 18 inches square, the top should be made only 17½ inches square.

For casting the slabs, choose a level area or working surface. The surface can be a garage floor, sidewalk, or wood platform. The base may also be sand; if so, lightly dampen the sand base before concreting. Spread a sheet of polyethylene across the base to simplify cleanup and to prevent the concrete from adhering to the surface.

The form is placed in position, oiled, and filled with concrete. A typical mix consists of 1 part cement to 2 parts sand to 2¼ parts of stone; if you want a mix with more coarse aggregate you could use 1:2½:3½. The mixture should be hardly more than earth damp and should require tamping to work into place.

In filling the form, tamp the concrete thoroughly and work the concrete into the corners of the form, spading along the edge with a trowel or similar flat tool or rod. The surface is then leveled with a straightedge.

Since the slab is cast with a stiff mix, the form can be lifted off immediately, cleaned, re-oiled, and placed in position for the next casting.

If there is any tendency for the concrete to slump or slough off after removal of the form, the quantity of water in the mix should be reduced for the next casting.

Immediately after casting, cover the slabs with damp burlap (or even damp newspapers). In curing the slabs, protect them from the sun and drying winds. Keep the slabs damp for three or four days. After three or four days the slabs can be lifted up and stacked on edge, but it is best to keep them damp (preferably covered with damp burlap) for at least a week after casting.

Surface Finishes

Various types of surfaces can be created on concrete slabs whether concrete is cast in place or precast. It is especially easy when working on small units in individual molds for precasting.

If a smooth surface is desired, the slabs may be finished off with a steel trowel as soon as the first water sheen has disappeared from the surface. (This may be about 1 to 1½ hours after casting, or somewhat less for a stiff mix.)

A nonslip finish can be obtained by finishing the concrete surface with a wood float rather than troweling.

A grooved surface can be made by pressing a ¼-inch diameter rod into the concrete surface immediately after it has been finished off. Circles, squares, or ovals can be made with household cans or plastic containers by impressing the can into the semi-hardened concrete surface.

Leaf impressions make interesting and decorative patterns. Press the leaves stem side down into freshly troweled concrete. Embed the leaf completely, but do not allow the mortar to cover the top of the leaf. Carefully remove the leaf after the concrete sets.

Another attractive finish for precast slabs is an exposed aggregate surface; see Chapter 11 for an explanation of techniques.

10. Paving with Precast, Garden Paths, and Other Things

Previous chapters dealt with normal construction of driveways, sidewalks, and patios. While these are common flatwork projects, and sometimes very necessary ones, there are other ways to pave a driveway, allow for parking, or build a walkway. And the interesting thing is that the result can be attractive as well as fun to build. You can do many things with concrete... here are a few.

Garden Paths

The charm and beauty of your garden will be enhanced by concrete paths, which will also simplify care and maintenance. The type of paving chosen depends on the garden layout, whether it is formal or informal, and on the area to be paved. There are several ways to build paths—whether cast-in-place or precast—all of which can be tackled with confidence by the homeowner.

Precast concrete slabs separated by grass and bordered by flowers create a pleasant entrance. (Portland Cement Assn.)

Large precast slabs used for a front sidewalk. (Portland Cement Assn.)

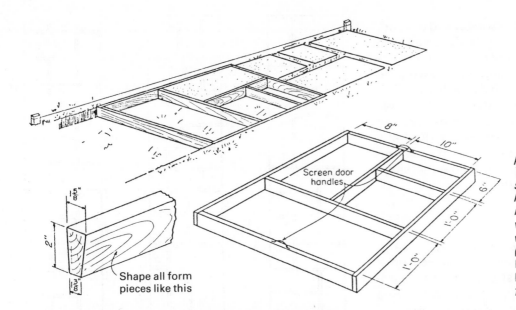

Shape all form pieces like this

Several slabs can be case in place at once by using this form. The form boards are beveled to simplify form removal. Handles attached to the forms make it easier to lift out. Reverse the form along the pathway to vary the pattern. (Reproduced from Concrete Improvements Around the Home, *Portland Cement Assn., 1965)*

A curved path around a flower bed built with precast concrete slabs. (Portland Cement Assn.)

crete mix. Place concrete in the form, tamp, and level with a straightedge. Then wood float or trowel, depending on the final surface desired. With this method the concrete is usually left in the form to harden for at least 36 hours before the form is stripped away (or, in this case, lifted out). Keep the concrete moist.

The form can be built of 1x3-inch lumber, which makes a slab about 2⅝ inches thick, or 2x4-inch boards can be tapered and assembled together as shown in the illustration. The boards are beveled for easy form removal. Handles can be attached to each side of the form so that the forms may be raised straight up without disturbing the freshly placed concrete. Reverse the form when repeating the pattern to get varied effects. Clean and oil the forms after each use.

Stepping stones for a garden path can be cast in place by digging away earth and sod in irregular shapes. (Reproduced from Concrete Improvements Around the Home, *Portland Cement Assn., 1965)*

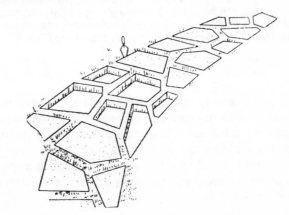

Precast Appearance

Several cast-in-place methods give the effect of precast slabs. Use of a gang form is especially effective for building garden paths, but could also be applied to building patios.

Instead of using a single mold, as described in Chapter 9 for precasting patio slabs, this method consists in having a gang form in which three or more paving slabs are cast together in place.

Remove the sod to the proper depth. Place a sand bed if needed. Lay in the gang form. Dampen the subgrade and oil the form. Use a rather stiff con-

The joints can be filled with strips of sod or sand. Stepping stones or garden walk slabs can also be cast in place by digging away earth to the shape desired. Stepping stones in a garden or lawn offer an inexpensive and picturesque pathway. The ground is first excavated to a uniform depth of two to three inches and the edges of the excavation trimmed. Before placing concrete, the bottom and the sides of the hole are damped down with water. Concreting, finishing, and curing follow the same steps outlined in earlier chapters.

Precast Slabs

The same type of precast slabs used in patio construction (Chapter 9), whether purchased from a building supply center or cast yourself, serve admirably in building garden paths and walks. See Chapter 9 for details on casting your own slabs.

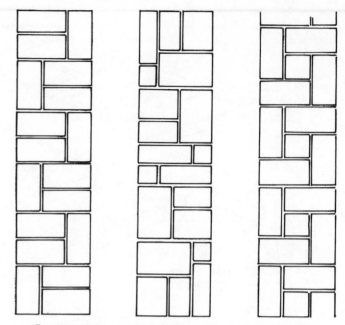

Precast slabs can be laid in a wide variety of patterns, and sizes and shapes can be varied to suit your taste.

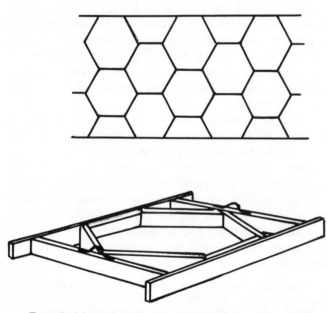

Top: A sidewalk built with full and half-size hexagons butted together. The half slabs can be molded in half size or a full slab can be cut with a trowel before the concrete hardens. Bottom: A simple form for precasting hexagonal slabs. Handles are added to simplify lifting the form.

If you prefer grass around the precast slabs, dig out the sod to the exact shape of the slab—round, square, or otherwise. Excavate to the depth of the slab if it will rest on soil, or deeper if you plan a 2-inch sand bed. Place the slab and tamp it gently into place. Leave a strip or area of grass between it and the next slab.

Perhaps you desire a slightly more formal walkway. In that case, lay out the walkway with stakes and mark the side of the path with taut string. Excavate the sod and soil to the depth of the slab, or deeper if a 2-inch sand bed is to be included. Level and compact the subbase. Dampen the subbase slightly to aid in compaction. Lay the precast slabs to the pattern desired. The slabs can be butted close together or different joint spacings can be used. Joints of ¼ or ½ inch can be filled with sand; wider joints can be filled with pea gravel or larger-sized stone or grass strip plantings. The joints can be random width, or if the same joint width is desired throughout, a strip of wood can be used to lay out and keep the joint spacing.

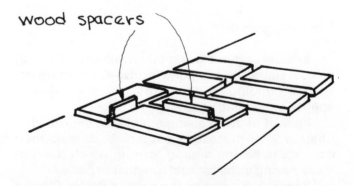

Strips of wood can be used to keep the joint spacing the same between slabs.

Concrete brick can also be used to build a walk following essentially the same techniques used for precast slabs. Excavate the walkway area to a depth of the brick plus 1½ to 2 inches for a sand bed. Cover the bottom with 1½ to 2 inches of sand; level, and compact. Lay the bricks flat on the sand. Gently tamp the brick into place. There are many patterns possible. Fill the joints between brick by spreading washed sand on the walk and sweeping it into the joints. Then wet the walk down with water to help settle the brick and pack the sand. Let the area dry before walking on it.

Crazy Paving

Yes, that is what it is called—crazy paving—and it can be used for patios or paths.

Quite often in home remodeling or other construction, old concrete walks or driveways need to be broken up and removed. You may have heard about people using these broken, irregular concrete slabs for garden walls; well, they can also be used in paving walks and patios. Confine your broken concrete to pieces taken from residential driveways and sidewalks; broken street paving is too thick and heavy for your use. Slabs may be as large as can be conveniently handled. If a slab is too large and heavy it can be broken to a more manageable size with a sledge hammer.

Admittedly not many people will be breaking up concrete sidewalks just when you happen to need them. However, quite often a neighbor or contractor will give you the broken concrete just for hauling it

Grouting exposed aggregate joints between broken slabs of concrete (Reproduced courtesy N. Z. Concrete Construction, New Zealand Portland Cement Assn., Wellington).

away. In other areas of the country these broken slabs are so popular that the concrete contractor has found that he can set a price on the slabs; the buyer has to haul them, or the contractor will deliver for a slight additional fee. Broken concrete is heavy; unless you have a pickup truck, or strong springs on your car, you may be ahead to have the contractor deliver it to your house. You may also find these broken concrete slabs up for grabs at a garage or moving sale, side-by-side with the bric-a-brac, and used furniture.

While the top surface of the broken slab is smooth or carries the finish of the original walk, the bottom side is rough and irregular. This means a little extra labor in excavating for the walk or patio and in laying the slabs. The sod and soil are excavated to a depth that accommodates the broken concrete and a 1½- to 2-inch sand bed. The bed is necessary to seat the irregular underside of the slab and to prevent it from rocking and breaking. Each slab needs to be tamped and seated securely. The slabs can also be placed on a lean mortar bedding for more permanency.

Being quite irregular in shape, the pieces when laid in a walk leave irregular joints and gaps between units. Try one of these attractive ways to fill the joints.

1. If the joints are kept relatively slim, sand can be swept in to fill.

2. Larger, irregular joints can be filled with soil and seeded with grass or strips of sod. Ground cover can also be planted in between the pieces.

3. And, for something a little different, a grouted exposed-aggregate joint filler for medium-wide joints between slabs. Aggregate sizes from pea gravel up to 1 or 1½ inches can be used.
   (a) After the slabs are laid, bedded, and leveled, the loose aggregate is spread in the joint spaces. Use a board or straightedge to ensure that the loose aggregate particles are level with the slab surface.
   (b) Mix up a grout composed of cement and washed sand in proportions of 1 part cement to 3 parts sand, and sufficient water to give the mix a fluid consistency. There must be enough water for the grout to flow easily through the voids in the aggregate. (Note: If, when your grout is in a bucket or other container, a considerable amount of sand settles

to the bottom within one minute, you have too much water in the mix.)

(c) Pour the grout mixture into the joints and let it flow in and around the aggregate. Then screed to slab level quickly with a straightedge or steel or magnesium float, as well as cleaning excess mortar off the slabs.

(d) When the mortar starts to harden—the water sheen has disappeared and the aggregate is not dislodged by brushing—take a stiff brush and carefully work the cement-sand mortar away from the top portion of the aggregate. A fine water spray, used in a limited amount, will help.

(e) You now have an attractive exposed aggregate joint finish between the broken slabs.

You haven't yet heard the craziest thing about crazy paving: the craze for broken concrete exceeds the supply in many cases, and some people are willing to make their own *new* broken concrete. So...here is how to make it.

Do not try to precast each separate piece in a different, irregular shape; the cost in time and labor would be too high. Simply pave a small area in concrete—cast-in-place concrete—and then cut the concrete into slabs of the desired shape before it has hardened.

The simplest way is to lay out your casting area in the same spot where you will be building your walk. Use a sand bed or sheet of polyethylene underneath. Set side and end forms and brace securely. A 2- to 2½-inch thick slab is adequate, so select form lumber accordingly. Set forms the width of the walk and set length at a manageable dimension for the amount of concrete you will be mixing.

You can also lay out a casting bed on any flat, level surface: driveway, sidewalk, or lawn, as long as it can be reserved for your work area for at least as week, and so long as formwork can be secured adequately to stand up to concreting pressures.

Use a mix similar to that used for precast patio units, but do not use coarse aggregate of too large a size. Large aggregates make cutting of the slab difficult. You may wish to restrict maximum size aggregate to pea gravel size; of course, a mix composed of 1 part portland cement to 3 parts of sand could be used also.

Dampen or oil forms before concreting. Place the concrete, compact it, screed, and finish. A wood float finish gives a gritty, nonslip surface. After the concrete sets, but before it hardens (about 1½ to

2 hours after casting) cut *all the way through* the slab with a trowel. Cut any shape you desire, but in marking out the irregular shape avoid pointy, acute angles that can break off easily while handling.

The fresh cut slab of concrete should be left undisturbed for 4 or 5 days to harden. But keep curing the concrete all that time by protecting the slab from rapid drying in the sun; keep the concrete moist.

Then this jigsaw puzzle of cut slabs can be lifted apart with a spade and each piece stacked in a convenient place until required. Use this new-old concrete for your sidewalk, with all its randomly placed slabs and irregular shapes, as described earlier for salvaged broken-up concrete.

Interlocking Precast Concrete Paving Blocks

Do you want a driveway or walk that looks different than the usual slab, can be bought from a local concrete block plant in small lots, and requires no special skill to install? Then you might want to consider the installation of interlocking precast concrete paving blocks (sometimes called "stones").

The blocks are produced in a wide variety of shapes and colors. Just a few of the styles are shown on the next page. Dimensions, of course, vary, as do thicknesses. Thickness of paving units range from about 2⅜ to 4 inches. A 2⅜-inch-thick block is commonly used for sidewalks, residential driveways, and bicycle paths. Since the shapes can be wavy rectangles, T- or I-shape, stars, interlocking hexagons, etc., there is no typical dimension; some example overall dimensions (nominal) are 9 x 4½ inches, 8⅝ x8⅝ inches, and 7⅞ x 6⅜ inches. A single unit 2⅜ inches thick can weigh between 8 and 10 pounds. Compressive strength of some of these mass produced blocks have tested as high as 8500 pounds per square inch. The top edges of each paving unit are beveled to enhance appearance and minimize chipping of the edges.

Preparation of the subbase and laying of paving blocks is similar to that used for other precast slabs. The sod, soil, and debris should be excavated to the necessary depth. Where drainage is a problem, or there are poor soil conditions, backfill with 2 to 10 inches of compacted granular fill (gravel, crushed stone, or slag). After this base has been compacted, cover with 2 inches of washed sand. Screed off the sand to the level required. In laying out excavations and fill, keep in mind that you want the paving block to settle down into the sand when it is tamped.

Place the paving blocks in the desired pattern as close together as possible; joint spacing should not

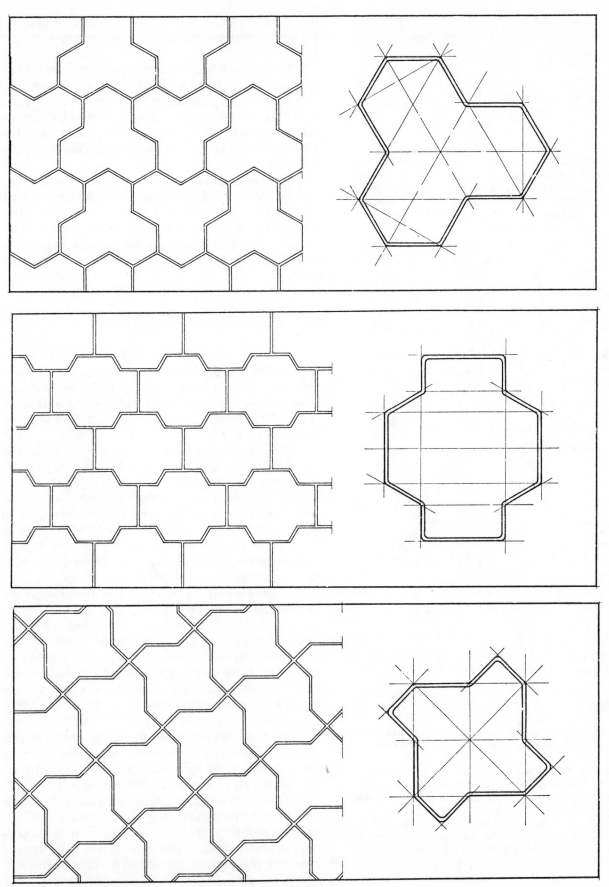

Precast paving blocks: the top edges are beveled for appearance and to reduce chipping and breakage of the corners. (Reproduced courtesy of Manual de Adoquines, Instituto Colombiano de Productores de Cemento, Medellin, Colombia, 1975)

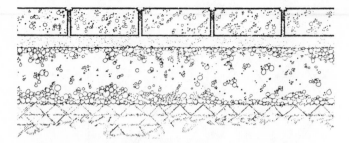

Precast paving stones should be laid on a sand sub-base. To assure adequate drainage and base for vehicular traffic, it is preferable that the sand layer be supported on a coarse granular base. (Reproduced courtesy of Manual de Adoquines, *Instituto Colombiano de Productores de Cemento, Medellin, Colombia, 1975)*

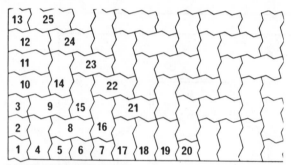

HERRINGBONE

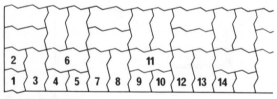

PARQUET

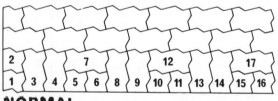

NORMAL

Patterns and order for laying interlocking precast concrete paving blocks. The herringbone pattern provides the best locking effect. Top: Herringbone pattern—begin in a corner; after the 8th block, you may continue at an angle of 45 degrees following the numbered laying sequence. Center: Parquet pattern—standard blocks are laid in pairs to develop double rows. Bottom: Normal or runner pattern—rows of blocks should be at right angles to the vehicle driving direction. (North American Stone Company Ltd. and KNR Concrete Systems Ltd., Ontario, Canada)

exceed ⅛ inch. It is important that joint spacing be kept consistent so that the pattern will remain constant.

Tamp down and level the units until the paving blocks are uniformly level, true to grade, and free of any movement. These can be tamped down with a hand tamper, although a mechanical plate vibrator will assure a better job. To finish the job, spread fine sand over the surface, sweeping the sand into the joints.

The pavements laid in this way can be used immediately after laying of the blocks. Another advantage of paving blocks, any style, over cast-in-place paving is that if there is ever underground work needed on utilities, sewer, etc., the blocks can be taken up and relaid without destroying the drive or walk.

Paving with Concrete and Grass

Developed in Europe and now available in the United States, a precast concrete waffle-like unit called the "Monoslab" or "Grass Paver" offers a new concept in environmental parking and walkways. The Grass Paver is a concrete turf and soil reinforcing grid.

The units lend themselves to a wide variety of uses around the home: overflow lawn-parking along the curb or on the berm, driveway shoulders or turnouts, footpaths and walks, patio, a base on which to stack firewood, rubbish container pad, or even a lawn pad on which to park a trailer, boat, or camper. The waffle units also make good facings for slopes, terraces, or banks along creeks or roads to prevent erosion and to stabilize the area.

As designed by landscape architect Paul Schraudenbach of Bavaria, the precast concrete units provide a load-bearing surface which spreads out the load of cars either parked or in motion. The configuration of the unit also provides pockets for soil so that the soil is not compacted under traffic and can retain a loose texture that is ideal for seed germination, healthy growth of grass, and free drainage.

The area "paved" with turf-grids drains easily. Water soaks through the grids down into the base and subsoil. Underground drainage is seldom required and storm water run-off is minimized.

Pedestrian walks and bicycle paths can be made by simply turning the turf-grids bottom side up.

The precast unit consists of concrete grids containing soil chambers and having regularly spaced concrete projections for wheel contact at the sur-

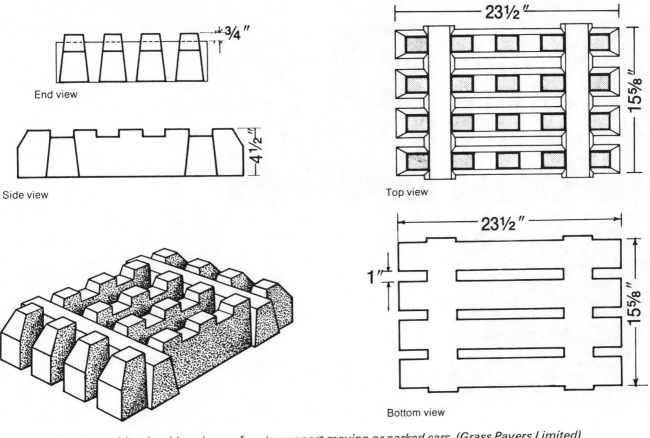

End view

Side view

Top view

Bottom view

Precast turf grids provide a load-bearing surface to support moving or parked cars. (Grass Pavers Limited)

face. A standard unit covers approximately 2½ square feet (23½ x 15⅝ inches), is 4½ inches thick, and weighs 92 pounds. Three of these grids will cover 8 square feet. The grids have a concrete compressive strength of 4000 pounds per square inch. The surface area is 75 percent grass or vegetation (or other fill) and 25 percent concrete, but the bearing area beneath is 88 percent concrete.

Preparation of a subbase is similar to that described earlier for driveways and walks. Excavate sod and soil, and where necessary backfill with a compacted layer of granular material (crushed stone, slag, gravel). Allow sufficient depth for ballast course, bedding layer, and concrete grid. Build to proper grade (see Chapter 6) and slope. In establishing grade, allow for the fact that the concrete grids will settle down into the sand bed when placed and loaded.

Where heavy loads (trucks) are expected, spread and tamp at least a 6-inch ballast course of coarse aggregate such as slag or crushed stone. Then a 1-inch layer of sand is tamped or rolled on top of the ballast. For light loads it is sufficient to spread and tamp or roll a 1-inch sand bed directly on the subgrade.

The turf grids can be laid in place by hand; they require no special skills, and can be put down in any weather.

After rolling or tamping the sand base lay the precast grids next to each other, keeping them "pencil thickness" apart. Where the area will be part

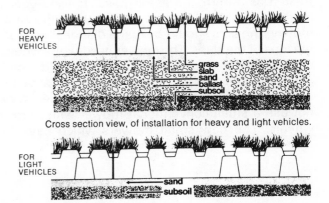

FOR HEAVY VEHICLES

grass
slab
sand
ballast
subsoil

Cross section view, of installation for heavy and light vehicles.

FOR LIGHT VEHICLES

sand
subsoil

The turf grid has pockets for soil and grass. Top: For heavy loads or where good drainage is required, spread and tamp a 1-inch sand bed on a 6-inch granular subbase. Bottom: For ordinary automobile loads and good drainage, spread and tamp a 1-inch sand bed directly on the subsoil. (Grass Pavers Limited)

Laying, filling and seeding precast concrete turf grids, called Grass Pavers or Monoslabs.

of the "lawn", use clean friable soil, or soil with a mixture of peat, spread over the grids. Level it off 1¼ inch below the top of the concrete, and sow grass seed. Fill further with ¼ inch of fine soil, and level with a hard broom. The final level should settle to 1 inch below the upper surface of the slabs. Always seed immediately after filling, while soil is still loose.

Freshly laid and sown turf-grids should be treated as a normal area of grass. Twice a year a mild fertilizer should be applied. When using a small mower, the cuts should be made diagonally across the tops of the slabs. The mower needs wide wheels so that they will not slip down into the grid. Large gang mowers can operate over such an area because they usually have large enough wheels; do not use a small conventional mower with normal wheels. The new small mowers which have no wheels but ride on a cushion of air (similar to a hovercraft) work well.

Where the concrete grids are used to stabilize terraces or slopes against erosion, ivy or similar ground cover can be planted to eliminate mowing on the slope or embankment. In laying the grids on an embankment, the long dimension should be laid parallel to the bank to minimize wash-out of the soil from the grid slots. If the slope exceeds 30 degrees, it is a good idea to stake half of the grids, in checkerboard fashion: every other course, every other grid. The stakes can be wood (¾ x 1½ x 24 inches) or bent scrap reinforcing rods formed into a steel "staple" (⅜-inch diameter rods with 20- to 24-inch legs).

For some areas—driveways, walks, street shoulders—rather than plant grass, fill the grids with pea gravel, marble or granite chips, sand, crushed red brick, or wood chips.

11. Color and Decorative Finishes

A molding system called Sculpcrete, developed in Australia, makes it possible to create fantastic and delicate designs in concrete wall panels. (Paul Ritter, PEER Institute, Perth, Western Australia, Australia)

Using the Sculpcrete process this artistic design was cut into the expanded polystyrene formwork, concrete was cast, and then the formwork removed. The concrete panel becomes an art form. (Reproduced from Humanizing Concrete, *Paul Ritter, PEER Institute Press, Perth, Western Australia, Australia, 1976)*

Open your eyes and imagination to form, texture, and color in concrete. Ours is an age of dynamic shapes and forms that colorfully express man's achievements. While concrete is a practical material, it can be molded into almost any shape to offer beauty and harmony—important requirements around the home.

Color...texture...geometric patterns...and other decorative finishes can be built into either precast concrete or cast-in-place concrete. Possible surfaces vary from glassy smooth to bold and rough, from nature's irregular shapes to sharp man-made geometric patterns. Concrete can also provide a variety of colors, from the icy blues of crystalline quartz, through the delicate pastels, to the flaming reds. Many different materials can be embedded in concrete. Textures can be varied. Patterns stamped into concrete create a surface that resembles stone or brick. And, finally, concrete harmonizes with and complements other materials such as brick and wood, as well as nature's shapes and colors: grass, ground covers, flowers, wood chips, and other surfaces.

While the emphasis in this chapter will be on use of color and decorative finishes in concrete flatwork—patios, walks, driveways, and other slabs—these same techniques and materials can also be adapted to concrete for other uses.

Color

Sidewalks, driveways, and patios take on new glamour when color is added.

There are a number of ways to achieve color in concrete: special aggregates, special cements, color pigmentation of the cement paste matrix around normal aggregate, and paints.

Hardened concrete can be colored by staining or painting. Color is created in fresh concrete by mixing (1) colored admixtures, (2) mineral pigments, or (3) colored cements into the basic fresh concrete mix. Another method is to apply dry-shake color to the fresh concrete surface before it hardens.

Dry-Shake or Dust-On Method

Dry-shake or dust-on color hardeners are ready-to-use products for coloring and hardening surfaces. A number of commercially prepared colored shakes are available. They are intergrinds of pigments, surface-conditioning agents, and portland cement combined with graded fine aggregates. They can be used to color and finish new concrete floors, patios, pool decks, walks, or driveways. They can be plastered on vertical surfaces of freshly placed concrete such as curbs or planters, but the method is really practical on only flatwork—horizontally cast slabs.

Color can be made intense because the color is concentrated near the surface, about the top 1/8 inch. It is not necessary to use color through the whole slab; the colored surface is wear-resistant. A surface made with a dry-shake color tends to be less porous than concrete finished without color, due to the density that comes with the material and to the finishing techniques used. Do not attempt to finish too much area at one time; timing of finishing is sensitive to temperature and humidity.

Applying dry-shake color to slabs. After the concrete has been screeded and darbied, and free water and excess moisture have evaporated from the surface, float the surface. Preliminary floating should be done before applying dry-shake material. This brings up enough moisture to combine with the dry material. Floating also removes ridges or depressions that might cause color variations.

Immediately following floating operations, spread the dry shake material evenly. In spreading the shake over the surface of the slab, do not throw it haphazardly; if too much color is applied in one spot, the result will be a nonuniformly colored surface. One good technique is to bend low over the slab and allow the material to sift through the fingers.

The shake is applied in two parts. About two-thirds of the total amount needed is spread over the surface once the water sheen has disappeared. In a few minutes this dry material will absorb some moisture from the plastic concrete, which should be floated. Then apply remainder of the shake, and float immediately. Keep in mind that the edges of slabs tend to dry out rapidly, and will usually be ready for the shake before the remainder of the slab.

Following the second floating operation, the slab should be trowelled. After the first trowelling wait for the concrete to set more completely, and then trowel again. The second trowelling improves the texture and produces a denser, harder surface.

At this point, draw a fine, soft-bristled broom over the surface to produce a roughened texture for a non-slip surface.

Cure dry-shake-colored slabs thoroughly.

Integral Color

Perhaps the "easiest" way to achieve colored concrete is to buy ready-mixed concrete that incorporates colored admixtures, pigments, or colored cements. However, colored ready-mixed concrete is not available from some producers due to the small demand for it, and it costs more. And that means you are using this more expensive concrete throughout the full depth of the slab.

You can mix your own colored concrete, but be forewarned that careful proportioning, measurement, and mixing are a must—and even then color may not be uniform between batches.

Colored admixtures. Admixtures can be used to color both flatwork and vertical concrete. Colored admixtures are blends of ingredients. They increase strength, improve workability, and disperse pigment and cement for easy placement and good color uniformity. The reduced water demand lessens color bleeding, particle accumulation at the surface, and efflorescence.

Pigments. Mineral oxide pigments added directly to concrete are used for color. They must be carefully selected and proportioned. Mineral pigments are usually less expensive than chemical admixtures mentioned above. If the pigments are not dispersed through the mix there is the possibility of color streaking.

The commonly used pigments and the colors they produce are shown in the table below.

Color	Pigment
Blue	Cobalt oxide Ultramarine blue Phthalocyanine blue
Brown	Brown iron oxide Raw and burnt umber
Buff, ivory, cream	Synthetic yellow iron oxide
Green	Chromium oxide Phthalocyanine green
Red	Red iron oxide
Gray or black	Black iron oxide Mineral black Carbon black

Pigments can be combined to achieve a wide variety of colors and shade. White cement should be used in the mix for light and medium shade colors. If ordinary gray portland cement is used, dark shades will result. The color will always be lighter after the concrete has dried out. The color of the aggregates will also have an influence on the overall color of the concrete surface.

Mixing. If you are mixing your own concrete, the amount of pigment used should not exceed 10 percent by weight of cement since larger quantities may reduce concrete strength. *The color is added by weight rather than volume.* The amount of color expressed by percentage is based on the weight of cement only; the weight of aggregate and sand does not enter the picture. A bag of cement weighs 94 pounds, so 8 percent color means 7½ pounds of color per bag of cement.

Where white portland cement and white aggregates are used, it is suggested that from ½ to 1 pound of color per bag of cement be used for pastels. Where gray cement and mortar sand are used, it is suggested that from 1 to 3 pounds of color per bag be used for pastels, and from 4 to 9 pounds color per bag of cement for deep tones. The water-cement ratio is also important. The more water used, the lighter the color of the concrete.

The best method for batching and mixing the concrete adds the color to the mix that contains the dry sand, aggregates, and the cement. Mix the batch for about 4 to 5 minutes. This enables the color to disperse evenly throughout the batch. Then add the specified amount of water, and mix the batch for an additional 4 minutes.

Placing. Place concrete when at a fairly stiff and buttery consistency. Slump should be 4 inches or less. The placing and finishing techniques needed for colored concrete are essentially the same as those for any concrete. Finishing should be done carefully and uniformly. Less trowelling is usually required; broom finish can just as easily be used. Curing is important; be sure to prevent moisture evaporation from the surface.

Intense colors are impractical because the entire thickness of concrete must be colored, and that becomes expensive. The cost of the coloring materials increases proportionately with slab thickness. Variations in water content and slump between batches can affect color uniformity, and variations in cement color can also affect the final color of the job.

Colored Cements

Most homeowners wanting colored concrete will use pigments for integral color or dry-shake colors, but colored cements are also available in three general types. The first is a standard portland cement that happens to have a light tint, maybe a light gray or slightly tan shade, colored by the specific raw materials used in the manufacture of the cement. The second is cement manufactured intentionally for color as a special cement. The third is a pigmented cement in which pigments were blended with portland cement, usually white cement. Color in these cements is reasonably consistent. Some good results can be obtained by combining colored cements with colored admixtures, but it costs more.

Curing Colored Concrete

Curing is the most important step when making colored concrete using the methods for dry-shake or integrally colored cement.

This is the one instance in concrete work where curing by water, fog mist, or wet burlap might cause problems because they can hurt color uniformity. Plastic sheets and waterproof paper also present problems in curing colored slabs because they cause uneven moisture distribution over the top surface of the concrete, resulting in a blotchy surface appearance. Water droplets condense on the bottom of the sheet or film. This water may drip off to form drops or puddles on the surface, or it may run down to form lines, causing discoloration. Likewise, spray-applied curing compounds may also affect color uniformity.

Because of these difficulties, colored concrete is often simply air-cured, and is therefore inferior in strength and surface hardness.

In general, colored concrete should be cured with material recommended by the manufacturer of the dry-shake hardener or coloring admixture. Use of a color-matched curing wax is the most popular method.

However, some of the tried-and-true methods for curing concrete can be used, if certain precautions are kept in mind. Moisture evaporation from the colored concrete surface can be prevented by any of these techniques.

1. Spray or swab liquid curing wax as soon as conditions permit.

2. Apply a clear sealing compound.

3. Cover surface with polyethylene film, *avoiding all wrinkles and air pockets,* and apply a sand layer on top of it to ensure continuous tight contact with the damp concrete surface. If you cannot cover the film with a sand layer, do not use this method.

4. Place 2 inches of clean sand over the entire surface and keep the sand wet.

5. Water curing, used only if the slab can be continuously flooded. Wet, clean burlap or cloth can be used if kept flat and smooth with no wrinkles and kept wet continuously. Do not allow alternate wetting and drying. Cover the entire surface.

Chemical Stains

Colored concrete can be obtained by applying chemical stains to cured concrete. Chemical stains are water solutions of metallic salts which penetrate and react with the concrete to produce insoluble color deposits in the pores of the concrete. They contain dilute acid to etch the concrete surface slightly so that the staining ingredients can penetrate deeper and more evenly.

The range of shades and colors is relatively limited on concrete stains. The colors available are usually black, green, reddish brown, and various shades of tan. Ferrous sulfate produces shades of orange to reddish brown; copper sulfate gives varying shades of green; commercial stains are available using other materials. (By the way, if you are using ferrous sulfate as the staining agent, do not be surprised that the concrete initially shows a greenish hue. After a short time this will change to a buff or brownish shade.)

The color produced is not a surface coating; it penetrates to a limited depth. The color will wear away to the extent that the concrete surface does. Stains will not hide defects or discoloration in the concrete, but they make the natural variations in texture less noticeable and reduce surface glare. For weathered, badly worn concrete, brown or black colors usually give the best results.

A little of the solution goes a long way. The stronger the solution, the deeper the shade produced. You may wish to mix up a small batch first, keeping track of your mix proportions. Stir the solution thoroughly until the powder has dissolved in the water. Use a soft brush to spread it over the

concrete in an area that is more or less away from the major "viewing area." Is it too light or too dark for your taste? The color can be made lighter or darker by reducing or increasing the strength of the solution. Also, a second coat of the same concentration will produce a darker shade.

Stains should be applied to hardened concrete which is at least 4 to 6 weeks old and free from all foreign matter. Stains work most effectively on relatively new concrete. The application instructions of the manufacturer should be followed carefully.

Whether staining new or old concrete, the surface must be free of oil, grease, paint and wax. New concrete probably needs nothing more than a good scrubbing with water to clean it off. On older concrete, remove surface grease with a solution of 1 pound of trisodium phosphate dissolved in 1 gallon of warm water, scrubbed into the surface with a stiff brush and afterward flushed clean with water.

The stain is usually applied in two thoroughly saturated coats with a soft brush. Work out the brush marks as much as possible because where they overlap the color will be deeper than elsewhere. Stain that collects in depressions will cause dark spots.

Anywhere from 4 to 5 hours to 4 to 5 days might elapse between applications of the first and second coats. It depends on the concrete, weather and ambient atmospheric conditions, and the stain used. It often takes 3 or 4 days for a stained concrete surface to reach its final color.

After staining has been thoroughly absorbed into the concrete, the surface is wet mopped or scrubbed to remove any residue. The true stain color will then be revealed.

Painting Concrete

Color applied to a hardened concrete surface is not as long-lasting as color which is incorporated into the mix or absorbed into the concrete. But, of the many techniques available for imparting color to concrete, painting is undoubtedly the easiest and probably the cheapest (especially if only initial costs are considered).

Paints provide the widest possible selection of colors, and many are sold for use on concrete. The types of paints are: cement-base paints, oil-base paints, varnish-base paints, lacquer-base paints, rubber-base paints, water-thinned paints, and epoxy paints. Some paints perform well under outside weather exposure and traffic, and some do not. Since there is such a variety, a book could be written on paints alone,* so only general guidelines are of-

*See, for example: *Book of Successful Painting* by Abel Banov and Marie-Jeanne Lytle, and *Paints and Coatings Handbook* by Abel Banov, Structures Publishing Company, Farmington, Michigan.

fered here. Whenever you are painting concrete, follow the manufacturer's instructions. Also check the label on the container to be sure that the paint was formulated for use on concrete.

The more common usage of the general types of paint are shown in the table below.

Common Usage of Paint on Concrete*

| General type of paint | Outside walls | | | | Inside walls (partitions) | Floors | | Surfaces exposed to water |
| | Above grade | | Below grade | | | | | |
	Exterior surface	Interior surface	Interior surface	Buried surface		On subgrade	Not on subgrade	
Oil-base	x							
Varnish-base		x			x		x	
Lacquer-base	x					x		x
Water-thinned	x	x	x					
Clear coatings	x							
Bituminous coatings				x				x
Cement base	x							x

*Reproduced courtesy of "Guide for Painting Concrete" and "Recommended Practice for the Application of Portland Cement Paint to Concrete Surfaces," American Concrete Institute.

Surface Preparation. Cement-base paints can be applied to damp concrete surfaces and to surfaces which will be damp frequently. With few exceptions, all other paints will serve best if the concrete is seasoned and dry when painted; blistering and peeling of the paint is likely to result if the concrete is damp (either exterior or interior). Any dirt, dust, grease, oil, or efflorescence should be removed from the concrete surface before painting. Dirt and dust can be removed by brushing, hosing, or scrubbing. Grease and oil may be washed off with solvent. Efflorescence can be removed by washing with dilute (1 part acid to 4 parts water) muriatic acid. Thorough rinsing of the surface with water should follow acid or solvent treatment. Before applying portland cement paint the concrete must be wetted thoroughly to control surface suction and aid in the hardening of the paint.

Application. Paints must be thoroughly stirred or agitated just before use to create a uniform distribution of pigment throughout the liquid. Any thinning of paint should follow the manufacturer's directions, and only with the recommended thinner.

It is best to paint when the temperature of the air and concrete is above 50 degrees Farenheit. Except for cement-base paints and water-thinned paints, the concrete surface should be dry.

The method of applying paint is usually not too important, provided the work is well executed. Brush, spray, and roller coatings are all good methods. Portland cement paints should be applied with stiff-bristled or whitewash brushes rather than ordinary paint brushes. Some lacquer-base paints are fast drying and are better adapted to apraying than to brushing.

It is seldom that one coat of any type of paint will provide the desired appearance and serviceability. Usually at least two coats are needed.

The drying time for the paint varies depending upon the type of paint and atmospheric conditions ...from 30 minutes for some lacquer-base and water-thinned paints up to 48 hours for some oil-base paints. Portland cement paint requires damp curing. On most jobs it is practical to sprinkle the painted surface two or three times a day for curing with the same fog spray as used for dampening the concrete. The curing should be started just as soon as the cement-base paint has hardened sufficiently not to be damaged by the spray, usually about 12 hours after application. Damp curing sould continue for at least 2 days.

Exposed Aggregate Finishes

This patio combines a coarse exposed aggregate finish in one section with slabs of tree rounds mounted in concrete in another. (Portland Cement Assn.)

Color and texture combine in these precast slabs set checkerboard fashion. The smooth, integrally colored slabs contrast with the exposed aggregate set into white cement. (Portland Cement Assn.)

Exposed aggregate is one of the prettiest and most popular decorative finishes for concrete slabs. Exposed aggregate finishes are not only attractive, but are rugged, nonslippery, and resistant to traffic and weather. It is a finish which can be applied to either cast-in-place or precast slabs.

Both the selection of the aggregates and the techniques in exposing them are important to the effect obtained.

Colorful, uniform-size gravel is recommended. Avoid flat or sliver-shaped particles or aggregates smaller than ¾ inch diameter; they do not bond well and easily become dislodged during the operation of exposing the aggregate. Select hard and sound aggregate. Native gravels are usually found in a range of browns. Some of the more attractive surfaces contain a large percentage of black stones. Stockpile the aggregate in a convenient location, to

be available immediately after the concrete is screeded and darbied. Keep the aggregate stockpile clean. The aggregate should be sprayed with water about the time the concrete arrives so that the surface of the stones will be damp when placed.

Making concrete and building the slab as the "base" for the exposed aggregate finish follows all the basic rules for building driveways, or walks, or patios as described in earlier chapters. Concrete with a maximum slump of 3 inches is preferred. Where the slab will be exposed to freezing and thawing cycles, use air-entrained concrete.

An exposed-aggregate-finish sidewalk blends with natural growth and landscaping. (Portland Cement Assn.)

Finishing Steps for Cast-in-Place or Precast

Immediately after the slab has been screeded and darbied, the selected aggregate should be scattered by hand and evenly distributed so that the entire surface is completely covered. Cover with a single layer of stones; if there are stones on top of stones the job of embedding the aggregate becomes more difficult.

The initial embedding of the aggregate is done by patting with a darby or the broad side of a piece of

Spreading aggregate: After the base concrete has been placed, screeded, and darbied, the selected aggregate is spread uniformly so that the surface is completely covered with a layer of stone. (Portland Cement Assn.)

Embed the aggregate by pressing and tapping it into the concrete with a wood hand float, a darby, or straight-edge. (Portland Cement Assn.)

A hand float is used for final embedding of the aggregate. The final appearance of the slab will look like a normal concrete slab after floating. (Portland Cement Assn.)

Exposing operations require flushing with water and brushing off the top cement paste surface. If aggregate is dislodged, delay the operation until the concrete sets further. Expose only the top surface of the aggregate; do not dig too deeply in brushing. Continue washing and brushing until there is no noticeable cement film left on the top surface of the aggregate. (Portland Cement Assn.)

2x4-inch lumber. After the aggregate is thoroughly embedded and as soon as the concrete will support the weight of a man on kneeboards, the surface should be hand floated using a magnesium float. Keep floating until all aggregate is entirely embedded just beneath the surface, and the mortar completely surrounds and slightly covers all aggregate, leaving no holes in the surface.

The time at which to expose the aggregate is critical. Shortly following floating, a surface retarder (a chemical solution which delays the set of the concrete surface) may be sprayed or brushed over the surface, following the manufacturer's recommendations. On small jobs around the home, retarders may not be necessary. However, it will give you several hours extra to finish exposing the aggregate, which may prove important in hot weather, when concrete tends to set faster.

Exposing operations should begin as soon as the concrete has set sufficiently to hold the aggregate firmly, generally ¾ to 1½ hours, and while the cement paste can still be removed by light brushing and flooding with water. Ideally the surface paste should be removed with a nylon brush and a full flow of water across the surface, using a water hose without a nozzle. When setting is too rapid, a stiff fiber broom and a nozzle (to allow greater water pressure) will have to be used, but care should be used to avoid dislodging the stones. The washing operation begins at a corner of the slab on the high side. Take only one pass; the paste that is washed away is carried forward in front of the broom to the edge of the slab. While washing and scrubbing, stay off the slab if at all possible because of the risk of breaking aggregate bond and dislodging the stones. If you must move out onto the slab, use knee-boards—and move them about gently; do not slide or twist the boards on the surface.

Brushing and washing should continue until the desired degree of exposure has been attained. Care must be taken to see that slightly more than half of the depth of aggregate remains embedded in the concrete. If one or two small spots are missed in the general wash, these can be exposed using a brush and pail of water.

About 2 to 4 hours after the aggregate has been exposed, it may be washed again and brushed lightly to remove any cloudy residue of cement film on the aggregate. The slab may also be washed with a solution of 1 part muriatic acid to 16 parts water. Following the acid wash, the entire slab should be flushed thoroughly with water. Weather and traffic will also remove the cement film.

Exposed aggregate slabs should be cured thoroughly. Employ normal curing practices, but take care that the method used does not stain the surface. Since only about one-half of each piece of aggregate is embedded in concrete, the slab must be well cured to develop full strength to attain bond between the paste and the exposed aggregate.

Conventionally Placed Concrete

In an alternate method, the aggregates are exposed in conventionally placed concrete. For success with this method the concrete mix should contain a high proportion of coarse aggregate to fine aggregate. The aggregate must be uniform in size, bright in color, closely packed, and properly distributed. The concrete slump must be kept low (3 inches or less) so that the coarse aggregate remains near the surface. Follow the usual procedures in placing, screeding, and darbying. Do not over-float since this could push the aggregate too deeply into the slab.

When the water sheen disappears from the surface you are ready to expose the aggregate. Follow the same procedures as described above.

Textured Finishes

Attractive patios can be built with large cobble-stones placed close together (above) or smaller stones spread more thinly (below). The latter patio effectively combines smooth slabs with exposed stones. (Portland Cement Assn.)

Cobblestone Surface

In some respects, a cobblestone surface looks like an enlarged edition of an exposed aggregate

Above: Cobblestones are embedded in the concrete slab with the flattest surface facing up. Below: Pattern of a walkway can be varied by selecting random shapes or elongated stones.

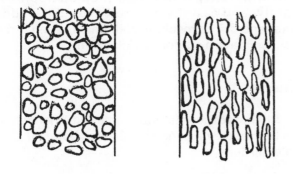

A wavy broom finish offers an attractive nonslip surface. (Portland Cement Assn.)

finish. A cobblestone surface can be achieved by embedding smooth, relatively flat stones 3-to-6 inches in diameter in the surface of the concrete. The stones are embedded with the flattest surface facing up. The slab is prepared in a normal manner; the large flat stones are placed by hand and pressed and tamped down into the concrete almost completely so that the stone surfaces are only a little higher than the concrete surface. Take care to use clean stones. Brush off any excess concrete which may squeeze up onto the faces of the stones. Then cure and clean the slab as described for exposed aggregate finishes.

Float, Trowel, and Broom Finishes

Interesting and functional textures can be created on concrete slabs just by using everyday finishing tools such as hand floats, trowels, and brooms— and that also means less expense.

Broomed finishes. Nonslip surfaces on driveways and walks are often made by drawing a broom at right angles across the slab, as discussed in Chapter 5.

A broomed texture can be applied in many ways: in straight lines, curved lines, wavy lines. If alternate slabs are cast with joints or wood dividers separating them, the brooming directions can be alternated so that the broomed lines are perpendicular to each other to give an attractive pattern that picks up shadows and highlights. Broomed precast slabs can be layed with similar alternating patterns.

After the concrete has set sufficiently so that the texture is not marred, the slab should be cured.

Swirl design. A nonskid surface, the swirl texture can be produced on a slab by using a magnesium or aluminum float or a steel finishing trowel. After the concrete surface has been struck off and darbied, a float is worked flat on the surface in a semicircular or fanlike motion. Patterns are made by using a series of uniform arcs or twists. A finer-textured swirl design is possible, from the same motion, by using a steel finishing trowel held flat. Moist curing of the slab is the final operation. Care must be taken to allow the concrete to set sufficiently so that the texture is not marred during curing.

A swirl texture can be produced with either a hand float or trowel. (Portland Cement Assn.)

A pitted texture is created by embedding rock salt in the surface and then dissolving it after the concrete hardens. These two rock-salt texture slabs are separated by a divider strip of regular concrete where some of the aggregate has been exposed. (Portland Cement Assn.)

Miscellaneous Textures

Pitted texture. Concrete surfaces can be given a rough, pitted appearance. After the concrete surface has been hand floated or trowelled, scatter large grains of rock salt over the surface. The salt is rolled or pressed into the surface so that only tops of the salt grains are exposed. After the concrete has hardened, the surface is washed and brushed, dissolving and dislodging the salt grains and leaving pits or holes scattered across the surface. Use large grain sizes and distribute the rock salt so that the resulting holes are about ¼ inch in diameter. Do not use this finish in climates subject to freezing weather! Water trapped in the holes and pits will freeze and expand and chip the concrete surface.

Dimpled concrete. Here is a technique that can be used if you are precasting your own walkway or patio slabs. The mold or form is prepared in the usual manner (see earlier chapters), but allow extra depth for the side forms to allow for a layer of stones. The bottom of the form is filled with coarse gravel, crushed stone, or slag—a single layer is usually sufficient. Cover the layer with a plastic film, allowing it to drape over and settle in to follow the surface. Cast the concrete on top of the plastic sheet. Screed off the top of the concrete. Since this will become the bottom of the slab, no further finishing is needed. Once the concrete has set and hardened sufficiently, the forms are removed and the slab is cured. The plastic

sheet gives a somewhat glossy finish to the concrete, with shadows and highlights provided by the dimples from the aggregate.

Wrinkled texture. This finish may be worth trying on precast slabs. Instead of filling the bottom of the mold with stones as was done for the dimpled look, wad up old newspapers. Line the bottom of the form with fairly firm wads, then dampen very slightly, and cast the concrete over the wadded newspapers. The resulting slab finish, when turned over, will have a random wrinkle effect.

Patterns

All kinds of geometric designs and other patterns can be created on concrete surfaces before the concrete hardens.

Random Flagstone

A random flagstone pattern can be scored or tooled into the surface. The scoring appears as recessed joints in the slab. The concrete should be scored or tooled after it has been screeded and darbied, and excess moisture has left the surface. This may be done by using a jointer or groover, or a piece of pipe bent to resemble an S-shaped jointing tool as used in masonry work. The tool is made of ½- or ¾-inch copper pipe about 18 inches long. One end of this tool is worked into the concrete to produce a scoring approximately ¾-inch wide and ⅜-inch deep. This should be done while the concrete is still plastic, allowing coarse aggregate to be pushed aside by the tool and embedded into the slab. The first jointing operation will leave burred edges. Following standard procedures for finishing, wait for excess moisture or water sheen to disappear and then hand-float the slab. Run the scoring tool again to smooth the joints. Then trowel the surface carefully. If a nonslip texture is desired, wait until the concrete has set sufficiently and then brush the slab lightly with a fine-bristled broom. The joints may be cleaned by brushing with a soft-bristled paint brush. Do not use water during the brushing operations. Finally, cure the concrete thoroughly.

Another way to create a flagstone design or random squared pattern is to temporarily embed 1-inch strips of 15-pound roofing felt in the concrete. After the usual operations of screeding, darbying, and floating are completed, the precut strips of roofing felt are laid flat on the surface in the pattern desired—random square, flagstone, or geometric. These are patted in and floated over. At this time a

A flagstone pattern can be tooled into the concrete surface. Above: A piece of copper tubing bent in a flat S-shape is used to tool fake joints in the surface. Below: After troweling the surface, the flagstone scoring can be cleaned with a soft bristled brush. (Portland Cement Assn.)

Leaf patterns in concrete. Embedding leaves in a fresh concrete surface. (Portland Cement Assn.)

color dry-shake may be applied, if desired. The slab is then finished in the usual manner. The strips are carefully removed after finishing and before curing. If color was used in exposed concrete, the joints under the roofing felt will be uncolored, adding to the "masonry" look.

Leaf Impressions

A leaf pattern can be used as a border to a patio slab, scattered across a precast slab, or used to decorate a garden walk. By using leaves from local trees, you can create a variety of patterns. Large, fresh, perfect leaves should be selected—picked at

the last possible moment before using. Immediately after the concrete has been floated and troweled, the leaves are pressed carefully, stem side down, into the freshly troweled concrete. Use a cement mason's finishing trowel to press the leaf into place. Take care to prevent cutting or tearing the leaves with the trowel edge. The leaves should be so completely embedded that they may be troweled over without dislodging them, but no mortar should be deposited over the leaves. For the best results the leaves should be left in place until the slab is cured, when they can be removed easily by pulling on the stem.

Circles and Squares

Cans, plastic containers, and large cookie cutters from the kitchen can create some unusual and attractive designs in a slab surface. Circles, ovals, squares of different sizes spaced out in planned or random design, or overlapping, add interest to any slab. These designs can be placed around the border, over the entire slab, or in alternate squares, and can be used in either precast or cast-in-place concrete.

After the concrete has been placed, struck off, darbied, floated, and steel troweled, the surface is ready to be given the imprint. Circles are a little easier to form than other shapes.

Using any number of circular-shaped cans of various sizes, press the open end of the can into the freshly troweled surface, starting with the largest

Above: Use a coffee can and other size cans to create circle patterns. (Portland Cement Assn.) Below: A closeup of interlocking circles in a concrete surface. (Concrete Construction Publications, Inc.)

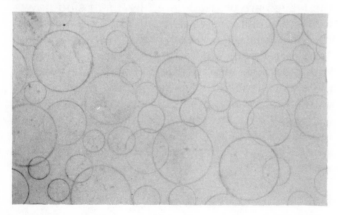

cans. Give the can a slight twist to ensure a good impression. Repeat to make a number of large circle impressions, then take the next-largest can and repeat the operation. The circles can be spaced in a regular pattern, randomly, or overlapping. Continue this operation until the desired number and sizes of circle impressions have been made.

If you are finishing a large area this way, you will need extra help. Although the concrete surface is quite hard after the final troweling, it can still be marked easily. But as the slab continues to harden, it becomes more difficult to make an impression. So, you have to work quickly and for this you might need help. By the way, be certain your helpers know the design pattern you want to follow.

Pattern Stamping

Stone, brick, and tile patterns can be cut into partially set concrete with special stamping tools. To make the concrete look even more like brick, the concrete can be colored integrally or by the dry-shake method.

Pattern stamping involves the use of patented aluminum tools to imprint the patterns on the freshly placed concrete.

The concrete should contain small coarse aggregate such as pea gravel, no larger than ¼ inch diameter aggregate. Too large a coarse aggregate creates difficulty in stamping the pattern. Concreting and finishing the slab follow usual procedures.

After the surface is floated or troweled, depending on finish texture desired, the stamping pads are placed on the slab. The stamping pads or platforms are placed on the slab and aligned accurately. By standing on the pad a normal-weight man will depress the grid design about 1 inch into the concrete. A handtamper can be used to ensure sufficient indentation. Progress is made across the slab by standing on one pad, lifting another, and placing it alongside. There are also single unit tools to stamp an impression for a single brick, and small tools to be used where the large ones will not fit and to dress edges and joints.

Timing is critical, since all stamping must be completed before the concrete sets too hard. The slab should be cured following standard procedures, as described earlier for plain or colored concrete.

A brick pattern in a concrete surface can be stamped in by using special metal stamping pads. One pad is placed next to the other so that the pattern is aligned. The design is stamped into the concrete by standing on the pad. A hand tamper can also be used to assure adequate indentation. (Portland Cement Assn.)

12. Fences and Garden Walls

Four different versions of concrete are used in the front entrance to this home. The lower wall (left) for the planter is 4x8x8-inch block. The grille/screen block at the entrance are 4x8x8-inch units. The sidewalk and steps are exposed aggregate concrete, while the porch slab is a plain troweled finish. (Portland Cement Assn.)

Concrete fences, garden walls, and embankments all add beauty to the home and grounds, but also serve functional purposes. The design should emphasize the more attractive features of your site. Also consider the neighboring properties; the construction should suit your home, but not conflict with the neighborhood.

Attractive construction is possible with cast-in-place concrete, concrete block, and precast concrete.

Footings

A wall is no stronger than its foundation. While a very low flower-bed wall (only one or two brick high) might rest directly on the soil, any wall or fence of substantial height should be supported on a concrete footing. In fact, almost all local building codes require footings—especially for fences and walls connected to a structure or higher than 3 or 4 feet, or those located on property lines. So—before build-

Self-supporting garden wall units with their own footings. The lacy, flower pattern is molded on both sides using the Sculpcrete process. (Reproduced from Humanizing Concrete, *Paul Ritter, PEER Institute Press, Perth, Western Australia, Australia, 1976)*

For a low, light wall built on firm soil, clay or hardpan, it may be satisfactory to excavate a trench the exact width of the footing and to use the sides of the excavation as forms for the footings. For added strength, reinforcing steel can be placed in the bottom of the footing.

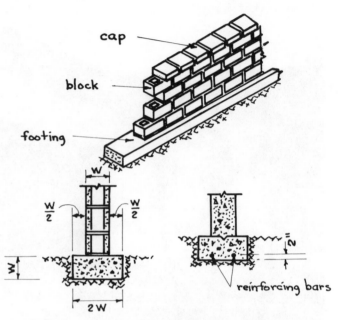

A garden wall should be supported on a footing. The footing width should be twice the wall width, and the footing depth should equal the wall width. If the footing is reinforced, keep the steel in the bottom one-third and at least 2 inches from the bottom.

A carefully excavated trench in firm soil can serve as the form for concrete footings without wood side forms. The reinforcing bars will be set on bar chairs or small pieces of concrete. (Reproduced from Manual of Concrete Inspection, *American Concrete Institute, 1975)*

ing that wall or fence, check with your local building authority to find out what the code specifies.

Codes generally set the criteria for footings based on wall height and width. A basic rule-of-thumb for residential footings erected on soils of average load-carrying capacity is that the footing width should be twice the wall width, and the footing depth should equal wall width. The wall should be centered on the footing if at all possible.

A low garden wall can usually be built on a shallow footing located just below the ground level. Depending on soil and drainage, it might rest on a gravel or crushed stone subbase. In localities where frost penetration is deep, the footing should be laid on a well-drained gravel bed to prevent moisture from collecting and freezing, thus eliminating or at least minimizing frost movement. In northern climates, the building codes will usually specify that foundations for high walls extend below the frost line. The depth of frost penetration varies around the country: for example, about 3 inches in parts of Tennessee, but as deep as 72 inches in northern Maine, Minnesota, and Montana. All soils can contain some moisture. When the soil freezes, the water in it expands. Then as it thaws, an equal amount of contraction occurs. This is known as "heaving." If footings are not placed below the frost line, the heaving can crack the foundations and footings and displace them, which in turn will damage the wall or abutting structures.

Typically, two or three #3 or #4 (⅜- or ½-inch diameter) reinforcing bars are sufficient. These are placed longitudinally the full length of the footing. If more than one length of bar is needed, lap the bars at least 15 inches and tie the lap together with small-gauge wire. The reinforcement should be located in the bottom third of the footing—a minimum of 2 inches above the bottom of the footing. To keep the bars at the correct height they are placed on chairs (made of steel wire, plastic, or even broken concrete pieces).

Often, the sides of the excavation cannot be used as the forms. In that case, dig the trench a little wider, but only as wide as you need to secure the working space to install formwork. Forms for footings can be built of nominal ¾-inch lumber. The forms should be able to withstand the pressure of the fresh concrete. Prefabricated steel forms also are available. The side forms should be securely staked and braced, and must be level. To keep the top of the forms from spreading as concrete is placed, nail 1x2-inch lumber (called spreaders) across the top at 3- to 4-foot intervals.

Proportion, mix, and place concrete as described in the chapter on slabs. Concrete should be placed as nearly as practical in its final position. As concrete is placed, spade it using a shovel or spading tool, just enough to compact it and to eliminate honeycombing (pockets in the concrete), especially next to the forms.

After the concrete is placed, a short piece of 2x4-inch lumber can be used to level the surface. Lay the 2x4 across the side forms and move it in a sawing motion as you proceed down the footing forms until a smooth, even surface is obtained. As you screed the concrete surface, remove the spreaders carefully, being sure the concrete is firm and will hold its shape.

Cure the concrete; wet burlap laid over the footing works well. The forms can ordinarily be removed after 2 or 3 days.

Fences and Brick Walls

Concrete fences offer strength and versatility, are also fireproof, and will not warp, rust, or rot. The desire for privacy, along with the rush to outdoor living that includes swimming pools, patios, and barbecues, has strengthened the demand for economical residential fencing.

There are two ways to build a fence: hire a concrete or masonry contractor, or build it yourself. These are some of the forms and materials and

methods: cast-in-place concrete, precast concrete, or concrete masonry. The fence can be solid and heavy-looking, or light and airy.

The concrete products and precast manufacturers offer a variety of units. The choices in block, post-and-panel, and palisade types of concrete fencing are many.

Blocks and Patterns

Standard or screen block fences, produced with lightweight or normal concrete on conventional block-making machinery, are available in every state. California pioneered in the construction of this kind of fence or wall.

Design possibilities are limited only by the imagination. There are probably 300 or more different block designs available, which can in turn be combined many varieties of ways.

Pattern through bond. When identical units are used, pattern is the feature to be developed; with

Concrete slump brick and screen block are combined for a fence that sets off this entryway. (National Concrete Masonry Assn.)

Slump block in a running bond pattern possesses a built-in adobe quality—ideal for a backyard fence. (National Concrete Masonry Assn.)

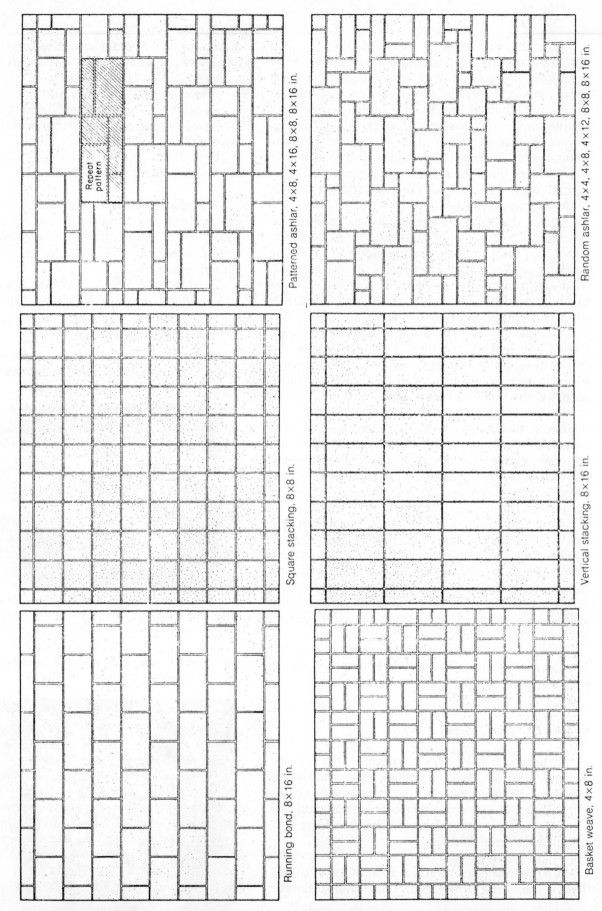

Patterned ashlar, 4×8, 4×16, 8×8, 8×16 in.

Random ashlar, 4×4, 4×8, 4×12, 8×8, 8×16 in.

Square stacking, 8×8 in.

Vertical stacking, 8×16 in.

Running bond, 8×16 in.

Basket weave, 4×8 in.

A sample of patterns available for concrete masonry walls, achieved by using various sizes of block and brick laid in various bond patterns. Many more patterns are possible. (Reproduced from Concrete Masonry Handbook, *Frank A. Randall, Jr. and William C. Panarese, Portland Cement Assn., 1976)*

standard block the choice of an appropriate bond is the easiest way to provide pattern. Assuming that normal running bond is too plain, stacked bonds (horizontal, vertical, square, or diagonal) are often used. Basket-weave and coursed and patterned ashlar are other patterns. Many of these can be produced from one or all of the standard sizes of block.

The ashlars (stonelike design achieved by using from one to five sizes of block in a specific design) may involve half-height or half-length units. Sometimes the scored block can be combined to give the same effect.

Also, you may create your own pattern. For example, a two-face effect can be achieved by alternately laying block at an angle to the line of the fence, so that the ends project to face in different directions. Similarly, a fence does not need to be straight; concrete block and brick can be made to follow any type of curve.

Joint patterns. Joints provide the next most apparent means for creating pattern. V-shaped joints give a neat appearance with sharp, contrasting shadow lines; normal concave joints give shadows which are much less pronounced; raked joints accentuate bond most of all. Weeping or extruded mortar, the direct opposite of raking, gives a highly rustic effect. To produce such a joint, the mortar squeezed out as the blocks are laid is not trimmed off, but left to harden in its extruded form. Combinations of flush and tooled joints are an interesting method of accenting either horizontal or vertical joints. For this approach, the joints to be emphasized are tooled, but the others are simply trimmed off with a trowel, rubbed with a carpet-covered wood float, and left flush.

Bas relief. An even greater emphasis on bas relief is possible through projected units. Projections allow interesting and economical shadow effects that break up large expanses of fence. Omitted block, the opposite of projections, offers pierced patterns at a cost savings. The inclusion of glass block, either at random or following a definite pattern, gives a similar but wind-tight effect.

Screen block. With screen block the pattern is inherent to the block itself. There are many blocks designed specifically for screening purposes (see Chapter 2), but there are also many types of regular masonry units which, when suitably arranged, provide a highly attractive fence. For instance, any type of hollow-cored block can offer an eye-catching pattern, on either side of the fence. One way is to lay the

A simple fence to build and yet one with dimensional emphasis. By offsetting units from the wall's plane, dramatic shadows enhance the backyard patio wall. (National Concrete Masonry Assn.)

An elegant backyard corner built with screen block. The toadstool (foreground) is concrete too. (National Concrete Masonry Assn.)

block in a running bond with the block openings facing out; another way is to alternate the units to lay horizontally, then vertically, with a gap in the middle. Similarly, chimney block on edge gives an inexpensive, attractive, and sturdy fence.

Screen block designs are legion. Almost every conceivable form and configuration can be molded: circles, stars, lattice, diagonals, crosses of every kind, letters of the alphabet, zig-zags and stylized butterflies. In combination with solid units, the scope is broadened even more.

Screen block of X-design gives a criss-cross pattern for a boundary wall. (National Concrete Masonry Assn.)

A simple screen block unit laid with open spaces between. (National Concrete Masonry Assn.)

A dramatic entryway is created using screen block with circular openings. (Portland Cement Assn.)

An open screen fence created by laying conventional concrete brick diagonally in a basketweave pattern. (Portland Cement Assn.)

Something different is achieved by setting blocks to protrude from the wall. This wall combines 8x8x16 inch, 4x8x16 inch, and lintel units. (National Concrete Masonry Assn.)

Combinations. Color, pattern, and texture also represent a combination that should satisfy any fence-builder with an eye toward beauty and economy. And...how about a combination of block and growing things? In southern climates where ice and snow are not a problem, partially cored units set crosswise in the wall can include recesses for small pockets of soil in which to plant flowers and creepers.

Apart from the aesthetic values, a combined solid-and-screen fence often gives precisely the correct ventilation and shadow arrangement you want. Too, there is another advantage: solid block can provide an effective acoustical barrier. Some city ordinances require a block fence around areas, such as parking lots, that abut residential property.

Mold Your Own Screen Block

Although you can buy a variety of screen block designs from local concrete products plants, you may want the fun of doing it on your own. One manufacturer (Decor Molds, West Caldwell, N.J.) has made it easy by developing small, durable reinforced fiberglass molds in which the homeowner can mold screen block, simulated stone, and coping block. There is, of course, some economy by doing it yourself; it is claimed that the screen block molded in your backyard will be about one-third the cost of ready-made commercial block.

In this do-it-yourself system, the screen block (design shown in accompanying illustrations) measures 12x12x3½ inches and weighs about 18 pounds. The block which resembles rustic stone walling measures 18x9x3½ inches and weighs approximately 35 pounds. The coping block, used for capping the top of a wall so water and snow will run off more readily, measures 12x5½x2 inches and weighs about 6 pounds.

Since you will be molding the units a few at a time at your own pace, you will want to mix the concrete in small batches. It is possible to use one of the prepackaged concrete or mortar mixes. A simple mix of 1 part cement to 3 parts sand, or 1 part cement to 2 parts sand and 1 part coarse aggregate, works fairly well. The aggregate must be kept small, no larger than pea gravel, since it will be tamped into the narrow screen block webs of the form. Mix

A rustic stone wall can be simulated by molding units in your own backyard or garage. The molds are made of fiberglass. Each concrete section measures 18x9x3½ inches and weighs about 35 pounds. (Decor Molds)

You can mold your own screen block. The block measures 12x12x3½ inch and weighs about 18 pounds. Foreground: fresh cast units. Background: A completed screen block fence. (Decor Molds)

Molding your own screen block. Top row (left to right): Mix concrete ingredients thoroughly. Add just enough water so that the mixture holds together when a ball is made with the hand. Place mold on a flat surface. Insert ejection plate. Fill the mold with the stiff concrete mix. Center row (left to right): Tamp the concrete firmly. Screed off excess concrete. Immediately turn over mold onto a plastic sheet. Bottom row (left to right): Eject freshly molded block right away by pressing on ejection plate through thumb holes. Remove ejection plate. The stiff concrete mix will stand by itself. Cure the units and allow them to dry out before installing in the fence. (Decor Molds)

sand and cement thoroughly, and add just enough water so that freshly molded block will hold together without slumping when the form is removed.

Blocks can be colored by adding pigments or coloring admixtures in powder form to the mix. And, of course, the blocks can also be painted. See Chapter 11 for further details on color.

The concrete is tamped into the mold, screeded off, and the mold turned upside down to eject the block. The dry, tamped concrete mixture will stand by itself. The easiest way to cure the blocks are to cover them with wet burlap.

The molds for the coping block and stone are simpler in form, but the units are cast in a similar manner.

Safety and Reinforcement

Whether the fence, garden wall, or patio screen is built of solid units, screen units, or a combination, it must be stable and safe. Follow your local building code. Ordinarily these walls are fairly low and are non-load-bearing; that is, they are not called on to support any more load than their own weight. Fences and garden walls should be designed and built to safely withstand wind loads of at least 5 pounds per square foot, and most city codes specify resistance to 20 pounds per square foot pressure. Since weather bureaus or the weather forecaster on radio or television seldom describe winds by pressure, but do report wind gust velocity: a wind-gust velocity of 40 miles per hour corresponds to a pressure of 5 lbs. per square foot; 80 miles per hour corresponds to a pressure of 20 lbs. per square foot.

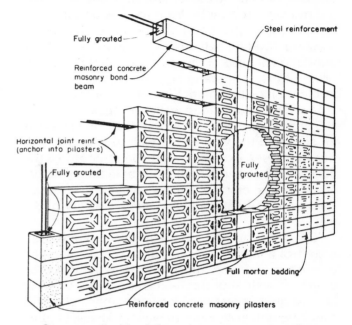

Screen wall with reinforced concrete masonry frame. (Reproduced courtesy of Concrete Masonry Handbook, *Frank A. Randall, Jr. and William C. Panarese, Portland Cement Association, 1976)*

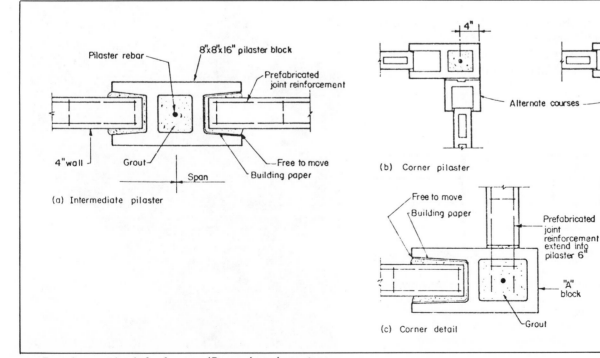

Framing methods for fences. (Reproduced courtesy of Concrete Masonry Handbook, *Frank A. Randall, Jr. and William C. Panarese, Portland Cement Association, 1976)*

Low masonry walls do not usually require reinforcement. High walls or fences should be reinforced both vertically and longitudinally. Some framing methods and reinforcement for screen walls and garden walls are shown here.

Control joints. Concrete masonry walls and fences are subject to movement, as in any structure. The movement may be from expansion and contraction due to changes in temperature, changes in moisture content, settlement or movement of the foundation or base, or movement of adjoining structures. Short, free-standing fences face no major problem since movement is not restrained to any large degree. But in long walls, or those abutting a slab or structure, the wall is restrained and stresses are set up in the wall as the wall tries to move. If these stresses exceed the tensile strength of the block, or the strength of the mortar in the joints between blocks, cracks occur to relieve the stresses. Cracks are unsightly and can affect stability of the wall.

Control joints and joint reinforcement are used to control and minimize cracking. The spacing and location of vertical control joints depend on the length of wall and architectural details. Control joints should generally be located at major changes in wall height, at changes in wall thickness, at junctions of walls and columns, near wall intersections, and at openings. Joints are ordinarily spaced at 20- to 25-feet intervals in long walls, and maximum spacing of control joints should in no case exceed 60 feet.

The control joints are usually located at vertical mortar joints to minimize cutting of masonry units. The joints should permit free movement, but have enough strength to resist required loads.

Tongue-and-groove concrete block for control joints are available in full- and half-length sizes. (Portland Cement Assn.)

A vertical control joint using conventional full- and half-size block. This joint, called the Michigan-type control joint, uses a strip of building paper to prevent the mortar from bonding to the block on one side of the joint. (Portland Cement Assn.)

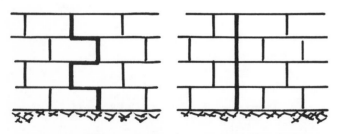

Control joints in a block wall can follow the normal bond (left) but this is not recommended. If a crack forms it may take the easy path through the middle of the block. The recommended vertical control joint (right) is formed by using half blocks to maintain the bond pattern.

There are a number of types of control joints that can be built into concrete masonry walls. There is a specially designed control joint block, with a tongue-and-groove design on the ends, which butts together. Or you can install a premolded rubber gasket in the joint. Another common method involves insertion of a strip of building paper in the end core of the block. Then the next block is laid. The core is filled with mortar. The mortar filling bonds to one block, but the paper prevents bond to the block on the other side of the control joint. Thus the joint permits movement, but also has strength. All of these control joints are first laid up in mortar just the same as with any other vertical mortar joint. After the mortar has partially set and is quite stiff, rake it out to a depth of about ¾ inch and apply a caulking compound or material.

Horizontal joint reinforcement will not eliminate cracks, but it minimizes large or unsightly cracks and also aids in developing lateral strength in the wall. The joint reinforcement should not extend through a control joint. While horizontal reinforcement could be laid in every course to reduce shrinkage cracking to a minimum, it would seldom be justified in garden walls. After the joint reinforcement is placed on top of the bare masonry course, the mortar is applied to cover the face sheets and joint reinforcement. Minimum mortar cover over the wire should be ⅝ inch.

Types of mortar. Masonry mortar is composed of cement-type materials and masonry sand mixed with sufficient water to give the mixture a workable consistency. The cement material can be either masonry cement or a portland-cement-and-lime combination. Masonry cements are easier for the homeowner to use because all the cement-type materials come premixed in the bag, and only sand need be added. A fairly common mix proportion (by volume) for masonry mortar is 1 part masonry cement to 2¼ to 3½ parts of clean, damp, masonry sand. There are also prepackaged masonry mortars available at most building supply centers; these are premixed and require only the addition of water.

Mortar can be mixed in small power mechanical mixers or by hand in a mortar box; mixing by machine is preferred. Mortar mixers of 4-to-7 cubic foot capacity can be rented from local equipment rental firms.

Mixing mortar. For machine mixing, first add a small amount of water to the drum; this prevents the mixture from caking up on the paddles. Initially add about three-fourths of the required water, one-half of the sand, and all of the masonry cement. Mix these together briefly. Then add the remainder of the sand and water until the desired consistency is obtained. After all ingredients are in the drum, mix for 3 to 5 minutes.

When hand mixing—and for small jobs around the home this is the more likely method—all the dry materials (masonry cement and sand) should be mixed thoroughly together with a hoe. Work from one end of a mortar box, or wheelbarrow, and then from the other. If you see streaks, continue mixing until everything is blended. Pull the material to one end of the box. Add about two-thirds to three-fourths of the required water and mix the dry material in with the hoe. Keep mixing until the batch is uniformly wet. Continue to add water carefully, with continued mixing, until the desired workability (consistency) is reached. Let the batch stand for about 5 minutes and then remix it thoroughly with the hoe.

Mortar should be used within 2½ hours after the original mixing and should not stand more than 1 hour without remixing; otherwise, it should be discarded. When the mortar needs to be remixed (retempered) on the mortar board, it should be done by adding water within a basin formed with the mortar and then reworking the mortar into the water.

Laying the wall. If you have any doubts about your ability to lay a masonry wall, or if the wall is complicated, high, long, or load-bearing—seriously consider calling in a mason or a masonry contractor. If you plan to do it yourself, and have never laid a wall, secure a basic instructional manual on bricklaying (blocklaying). Here are the basic guidelines.

Tools needed include assorted trowels and jointers, level or plumb rule, brick hammer and brick set, steel square, line, retractable steel tape, and folding rule. You will also need a mortar board: a 2 or 3 foot square platform to hold freshly mixed mortar. A story pole can also prove very handy.

Loose aggregate, dirt, and debris should be cleaned off the top of the footing so that the mortar will bond to the foundation. Make sure that the foundation is aligned vertically and horizontally. Any corrections must be made before starting masonry construction.

Snap a chalk line to mark the outside edge of the wall on the footings for alignment of the units. Lay out corners, intersections, and openings to establish proper alignment and joint locations.

Lay the corner units first, then level and align. The first course should be laid with great care, as it will assist in laying succeeding courses and in building a straight, plumb, wall. When three or four units

103

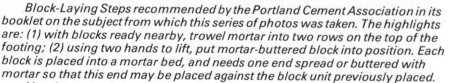

Block-Laying Steps recommended by the Portland Cement Association in its booklet on the subject from which this series of photos was taken. The highlights are: (1) with blocks ready nearby, trowel mortar into two rows on the top of the footing; (2) using two hands to lift, put mortar-buttered block into position. Each block is placed into a mortar bed, and needs one end spread or buttered with mortar so that this end may be placed against the block unit previously placed.

Use the mason's level (3) to make sure blocks are level, tap down (4) blocks at points where block remains slightly above the stretched mason's line cord. Blocks at building corners are laid up first to provide line stretching clips a fastening base. The level is in almost constant use to check face straightness (5) as well as level and plumb. With one edge of the trowel (6), excess mortar is bladed off.

When mortar begins hardening, the joint tool is used to give each mortar joint a slight indenting pressure and a smooth surface (7). When proper course height is reached, masonry anchor bolts (8) or metal clips are set and block core holes fill with mortar. And in (9) a final step, outer surface of the wall is given a cement coat (parging).

have been laid, use a level to check them for alignment, grade and plumbness.

In laying the first course, a full mortar bed should be placed on the foundation for the full thickness of the wall.

The corners are usually laid up about four courses high, stepping back each course by one-half a unit length. Each course is carefully plumbed and checked for alignment both vertically and horizontally. Measure with a story pole at the corners to locate the top of each masonry course.

The wall is then filled in between the stepped corners. A string line is stretched tightly from corner to corner to guide you in laying the top outside edge of each block for each course at proper line and grade. All aligning and plumbing of each unit to final position must be done while the mortar is soft. Any adjustments made after the mortar has stiffened will break the mortar bond.

Be careful not to spread the mortar too far ahead of the block being laid, or it becomes stiff before the units are laid. The masonry units must be set in soft plastic mortar for proper embedment and bond. As each block is laid, the excess mortar extruding from the joints should be cut off cleanly with a trowel. This mortar can be thrown back on the mortar board and reworked into the fresh mortar on the board. Discard any mortar which falls on the ground.

Try not to smear mortar on the block faces, since smears are difficult to remove. Allow any mortar droppings that stick to the face of the wall to dry before removal. Brushing will remove most of the mortar adhering to the surface.

Fence Posts

There are other types of fences. Concrete posts are used instead of traditional fencing in some cases.

Since cross-sections of posts are relatively small, the aggregate needs to be kept fine and the reinforcement light. Surface appearance is also important. Maximum size aggregate should not exceed ¼ or ⅜ inch; a mortar mix of sand and cement would also work. Typical proportions call for 1 part cement to 4 parts aggregate. The mix should be relatively stiff, yet plastic, so that it can be easily worked under and around the reinforcement with a stick. The concrete must be compacted thoroughly. White portland cement can be used to give the concrete a lighter color.

The posts can be cast in single or gang molds, and to a variety of shapes. A simple gang mold is il-

lustrated, as well as a mold for a single unit. Molds should be assembled on a flat surface. The bottom and sides of the mold should be coated with light oil before each use.

While longitudinal concrete plank or beams can run between posts built with slots or inserts to receive them, a simpler fence combines concrete posts with wood slats or boards which are bolted on. It is easy to cast bolt holes in the concrete post. Holes cast in posts should be ⅛ inch greater in diameter than the bolts or tubes they are to accommodate. The holes can be formed by dowels or pipe that pass across the mold from one side to the other. They should be coated with oil before concreting, and their removal will be facilitated if the rod is turned to break the bond before the final set of the concrete.

Square concrete rails mounted in concrete posts set off a home, without overpowering the landscaping. (Portland Cement Assn.)

A simple fence built with concrete posts and flat wood rails (lumber) spanning between. (Asociacion Venezolana de Productores de Cementos, Caracas, Venezuela)

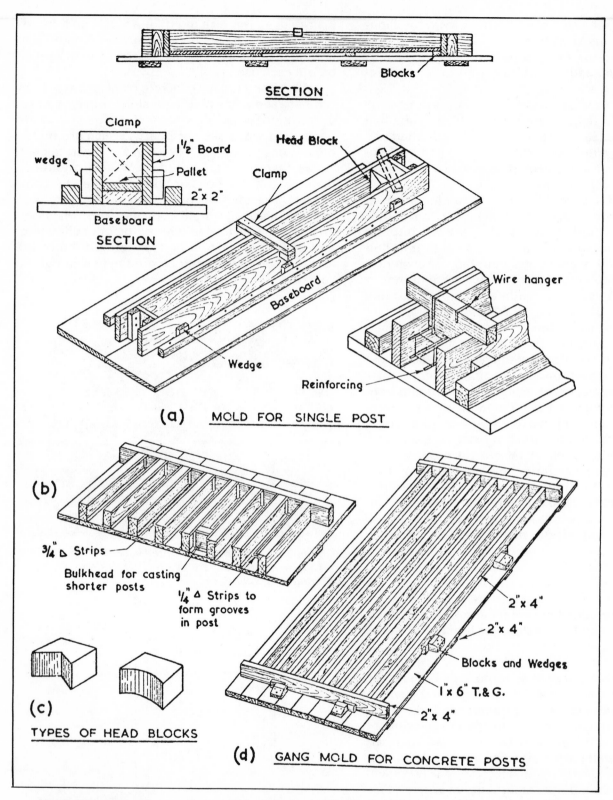

SECTION

Clamp

wedge

$1\frac{1}{2}''$ Board

Pallet

$2'' \times 2''$

Baseboard

SECTION

Blocks

Head Block

Clamp

Baseboard

Wire hanger

Reinforcing

Wedge

(a) MOLD FOR SINGLE POST

(b)

$\frac{3}{4}''$ △ Strips

Bulkhead for casting shorter posts

$\frac{1}{4}''$ △ Strips to form grooves in post

$2'' \times 4''$

$2'' \times 4''$

Blocks and Wedges

$1'' \times 6''$ T.& G.

$2'' \times 4''$

(c)

TYPES OF HEAD BLOCKS

(d) GANG MOLD FOR CONCRETE POSTS

Molds for concrete fence posts—single or gang molds. (Portland Cement Institute, Johannesburg, South Africa)

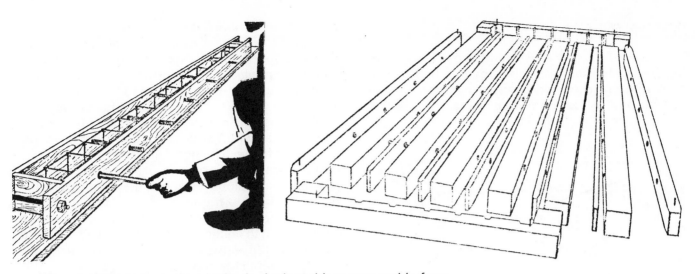

When molding fence posts, whether in single molds or gang molds, form bolt holes for attaching wood rails or boards by installing dowels or pipe through the form. (Instituto Mexicana Cemento y del Concreto)

Reinforcement for a post usually consists of four #3 (⅜ inch) reinforcing bars, placed near each corner. The correct placement of the reinforcement in a mold is essential to adequate cover for protection of the steel from corrosion, and to avoid spalling of the concrete. A reinforcing bar should not be any nearer to the surface than ¾-inch. It may help to tie the four bars with light wire into a simple cage. One technique of holding reinforcement in position during casting consists of suspending the bars by wires from a saddle over the mold; when screeding the concrete, remove the saddle and finish the concrete surface. Another approach is to rest the cage on concrete spacers which, in turn, become part of the post. Be sure not to dislocate the reinforcing bars during placing of the concrete.

For a pleasing appearance, quarter-round molding or triangular strips can be placed in the corner of the forms. This gives a beveled edge to the concrete post.

When placing concrete, spade it well along the sides of the forms and around all corners. Tamp the concrete—without dislocating the reinforcement. You might even tap the form lightly with a hammer to release any trapped air bubbles. Leave the post forms in place for at least 24 hours.

After stripping the forms, fill any small bugholes in the concrete surface by working a cement-sand mortar into the surface with a wood float. Moist-cure the post for 5 to 7 days.

Embankments and Terraces

Tree Wells

Sometimes in grading a lawn you must change the existing slope so that trees already on the site will be either above or below the new grade. The trees can be transplanted, but this is expensive and the trees might not survive. If the trees are below grade, they will die unless a well is left around them.

Circular walls of stone, brick, and block can be built around the well. The wall is a small retaining wall. Also a sloped embankment is safer when small children are around. The embankment can be held in place with turf-grids or concrete block.

Ordinary concrete block are used to form a tree well. Ground cover can be planted in the core openings. (National Concrete Masonry Assn.)

Retaining Walls

V-block retaining wall. An attractive low retaining wall or border for a raised garden bed can be built with precast V-block (see illustration). These can be purchased from local concrete product plants.

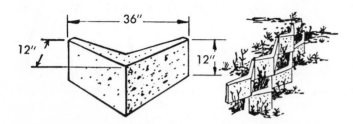

A picturesque low retaining wall or flower terrace can be built with precast concrete V-block. (Aero Welding and Manufacturing Company, Inc.)

Turf-grid revetments. Steep banks along creeks or road cuts can give the homeowner problems. Many times grass or ground cover will not hold and the bank soon erodes away. To solve the problem, use special concrete revetment block. The perforated waffle-grid units described in Chapter 10 work well in such situations.

Banks can be stabilized by laying precast concrete turf grids. The grids should be anchored with stakes through every second or third slab. Then grass or ground cover is planted. (American Concrete Institute)

Crib wall units. Crib units offer both economy and attractiveness when a retaining wall is needed for a steep embankment. Crib units were designed basically for embankments along road or railway cuts, but the smaller units are quite suitable for home use. Crib units consist of stretcher, header, and closer precast units to form an open box or grid. Many concrete products plants make them. Crib walls may be vertical or battered (see glossary). They are usually constructed with a batter of 1:75 and the base width or depth ranges from 50 to 100 percent of the wall height. The first course must be laid accurately in order to provide the finished alignment and slope chosen. It may be desirable to place a concrete foundation (see drawing). The open box forms making up the crib are filled with stone or soil to create an economical wall. The grid can be planted with ground cover to make an attractive bank, with little maintenance required.

CRIB RETAINING WALL

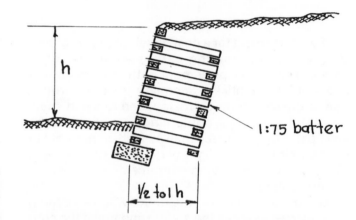

A crib is built of precast concrete units and offers a way to hold embankments in place, and at the same time allows ground cover to be planted and grow in the open grids.

Small crib wall units being placed along an embankment (Reproduced courtesy N.Z. Concrete Construction, New Zealand Portland Cement Assn., Wellington).

13. Something Different & Beautiful

Benches, Sundials, Pottery, Garden Pools, Bookends, Canoes

Concrete can be fun and artistic as well as being useful in foundations, buildings, bridges, and pavement. Concrete and mortar projects offer a variety of ornamental effects. You can buy or cast small concrete statues of squirrels and frogs, bird baths, and many other shapes. You can cast flower boxes and bookends; almost anything can be molded out of concrete.

Garden furniture comes in many styles and sizes. Some benches are made entirely of concrete; others utilize wood for seats or back. (Aero Welding and Manufacturing Company, Inc. and Portland Cement Institute, Johannesburg, South Africa)

Bookends

Perhaps you might like to start with a small project to see if you enjoy working with concrete. Why not try concrete bookends? Build a simple mold: a box 4 inches wide, 8 inches long, and 4 inches deep. Coat all wood pieces with two coats of shellac. Assemble and nail but keep nail heads protruding out

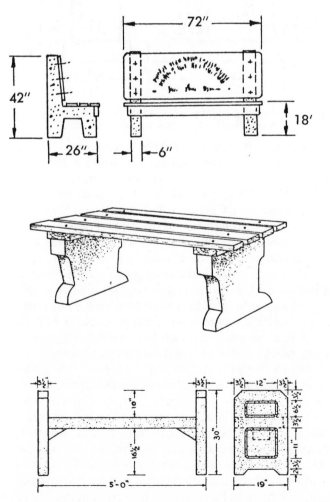

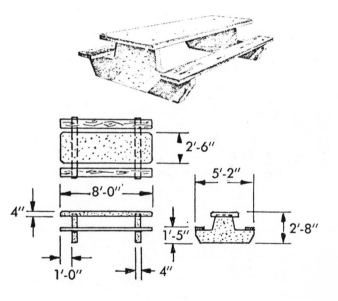

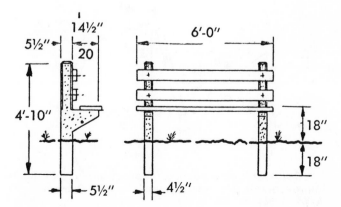

Practical and Decorative Concrete

so that the form can be stripped easily after casting. Coat the inside of the mold with light oil.

Use the same stiff concrete mix described for use in pottery (see later in this chapter). White cement can be used for lighter tone concrete. Tamp the concrete into the mold. If you have some attractive stones, tamp them into the concrete surface to a little more than half-depth. Otherwise, trowel the surface smooth.

Strip the mold and allow the concrete to dry out. Once the block is perfectly dry, glue a piece of felt or rubber sheet to the base so that the block won't scratch the bookshelf.

Concrete Benches

Depending on locality, you can buy various styles of garden furniture and benches from precast concrete producers. But in case you want to try your hand at building some, there are some simple yet ar-

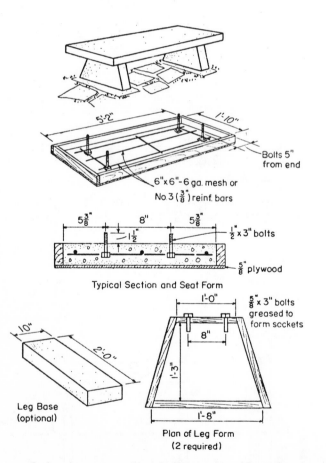

A simple concrete bench. (Reproduced from Concrete Improvements Around the Home, *Portland Cement Assn., 1965)*

tistic designs that can be made with a minimum amount of skill and expense.

Here are some step-by-step instructions for building one of the many possible designs. A detailed drawing of end supports for the bench pedestals is shown in Figure 13.1, and details of each piece used in making the wood mold for these end supports are shown in Figure 13.2. The assembled mold, ready for placing concrete, is shown in Figure 13.3.

The first thing to do is take a 1-inch board (or plywood), 14 inches wide by 16 inches long, and cut

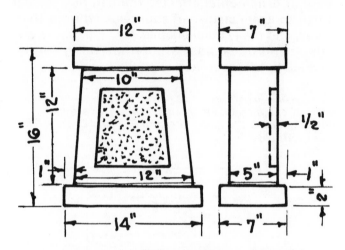

Fig. 13-1 *Bench pedestal. The indented area can be tooled with a hammer and chisel.*

it to shape A in Figure 13.2. This is to be used for the bottom of the mold as shown in Figure 13.3. Now cut another piece of board (also 1-inch thick) into the shape of B in Figure 13.2, and nail it down at its proper position on piece A. The next piece (C in Figure 13.2) forms the recessed panel; this panel is only ½ inch deep, so cut it from ½-inch board. Nail this securely in position, as shown in Figure 13.3, onto piece B. Be sure to bevel the edges of pieces B and C as indicated by the dotted lines, because this will make it easier to release the mold from the concrete once it is set. Now proceed to make pieces D, E, F, G, H, I, J, K, L, and M—all from 1-inch-thick material— being careful to follow the dimensions shown.

When the pieces are cut, assemble them as shown in Figure 13.3. Before assembling the mold, shellac each piece thoroughly on both sides as well as on the ends. This will prevent the wood mold from absorbing moisture and thus prevent any tendency of the mold to warp or buckle. Use as few nails as possible in fastening the pieces to one another. When the concrete is hardened, the form or mold will have to be removed from it, and the fewer the nails used, the easier the form can be stripped from the cast concrete.

110

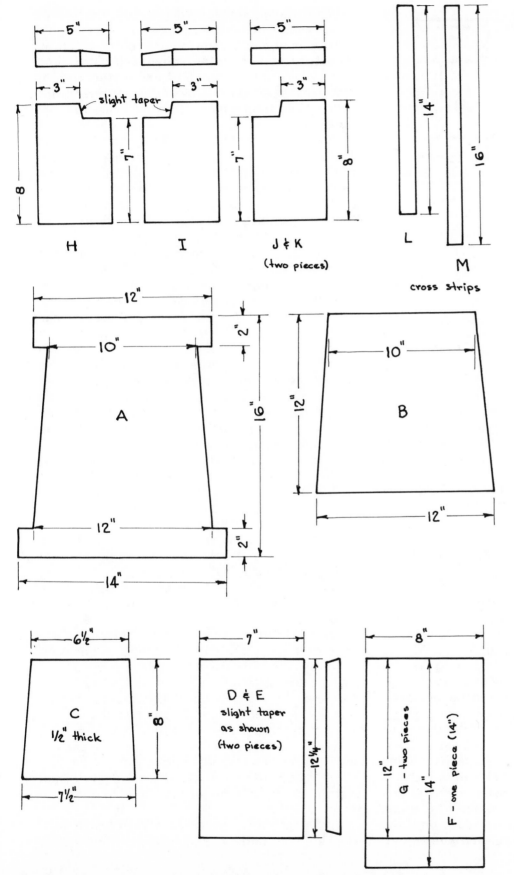

Fig. 13-2 *Wood pieces required to build a mold for a bench pedestal.*

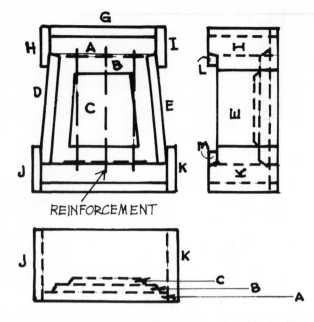

REINFORCEMENT

Fig. 13-3 *Assembled mold for bench pedestal. The parts A-M shown in Figure 13-2 are indicated here in the assembly. The reinforcement can be welded wire fabric.*

After assembling the pieces, as shown in Figure 13.3, inspect the mold to see whether the joints fit together well. If not, they can be filled with putty or plaster of Paris, being careful to keep everything square and true.

After having trued the mold up, the inside of it should again be shellacked and, when thoroughly dry, a thin coat of fairly thick oil should be given to all parts of the mold which will come in contact with the wet concrete. The mold is now ready to be filled with the concrete mixture.

You can use one of the prepackaged concrete mixes. Or, if you start with your own mix, use a typical mix of 1 part portland cement, 2 parts of clean sharp sand, and 2 parts of gravel pebbles (pea gravel) ranging in size from ¼ to ½ inch. For this small a project, you are likely to be mixing the concrete by hand in a wheelbarrow or a mortar box. Spread the sand out and then spread the cement on the sand. Thoroughly mix the cement and sand together until it has a uniform color. Water should be added carefully, and the mass turned over and over by means of a shovel or hoe until it reaches a uniform consistency equivalent to fairly thick putty. To this mortar add the stone or gravel which has previously been dampened with water. The whole mass should be mixed or turned over until the coarse aggregate is thoroughly coated with mortar.

The concrete should be deposited in the form or mold as soon after mixing as possible. Fill the mold with a shovel, carefully pushing the concrete into all of the corners. Tamp or tap it down well with the end of a piece of board, or reinforcing steel rod, or trowel. If the concrete has been properly mixed, this tamping will bring to the surface of the mass a slight water sheen.

The mold should first be about half filled and then a piece of welded wire fabric or other metal fabric should be placed into the form as reinforcement (see the dotted lines in the plan view of the assembled mold in Figure 13.3). This could be made of 4x12—W2.1xW0.9 welded wire fabric (also called 4x12—8x12).* Metal lath might also be used.

After having placed the reinforcement, continue to deposit the concrete, and tamp it down until it is level with the top of the sides D and E of the mold. Screed and float this surface level. Then, take the cross strips L and M, shown in Figure 13.2, and nail them to the top of the mold and against the end pieces H, I and J, K, as shown by the dotted lines in the side elevation in Figure 13.3. These cross pieces not only act as a form for the edges "a" and "b" of the pedestal, as shown in Figure 13.1, but they also brace the sides of the form, and prevent them from spreading apart due to the weight of the plastic concrete pushing against them.

After having nailed these pieces in place, fill the portion of the mold thus formed with concrete, until flush with the top of the strips and the end pieces G and F. Tamp the concrete down and smooth the surface off. The filling of the mold is now complete, and it should not be disturbed for at least 24 hours. It is preferable to keep the whole unit covered with wet burlap for at least 2 days. After having set for 2 days, the cast can be removed from the mold and set aside to cure further. Keep the unit damp for at least 7 days.

Be careful when removing the mold not to damage it or the cast, since the same mold will be used for the second pedestal. In removing the mold from the cast, first detach the cross strips L and M; then the pieces H, I, J, and K; then the end pieces F and G; next the side pieces E and D; and then the bottom piece, composed of pieces A, B, and C. Before using the mold again, it should be thoroughly cleaned of any concrete particles adhering to it. After cleaning it well, oil the inside and assemble as before. Then cast the other pedestal for the bench in the same manner.

The next step is to cast the slab or bench seat. This is 5 feet long by 18 inches wide by 3 inches

---

*AUTHOR'S NOTE: The first designation, 4x12, indicates the space between wires in inches; the second, 8x12 or W2.1xW0.9, refers to gauge of the wire.

thick. The form or mold for this is nothing more or less than an oblong box, having a bottom 5 feet long by 18 inches wide, and four sides that are each 3 inches high (see Figure 13.4). If you want the top edges of the slab beveled off, a triangular strip of wood can be nailed along the bottom edges of the mold (see the cross-section of the slab mold in Figure 13.4). The bench seat should be reinforced with the same size of wire fabric as used in the pedestal, or by three #3 (⅜ inch) rods spaced 6 inches apart. The reinforcing steel should extend within 3 inches of all four sides, and should be placed about ½ inch from the under surface of the slab.

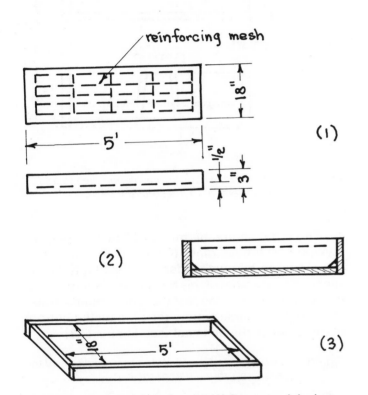

reinforcing mesh

(1)

(2)

(3)

Fig. 13-4 *Bench slab and mold. (1) Concrete slab showing reinforcing. Reinforcement is placed in the bottom part of the slab. (2) Beveled edges can be formed on the top edges of the slab by inserting triangular pieces of wood in the mold. In this case you cast the slab upside down, and the reinforcing is placed near the top of the concrete. Turn the slab right side up when placing it on the pedestal. (3) The form is a simple box.*

It would be a good idea to clearly mark the under surface of the slab, so you know to which surface the reinforcing will be closest once the slab is cast; it is important when placing the slab on the pedestal to always have the reinforcing nearest to the under-

side of the seat. That is where it is needed to resist tension.

Shellac and oil the inside of the mold and proceed to fill it with a mixture of concrete composed of the same material as was used for the pedestals. First fill the mold to a depth of ½ inch, then lay in the reinforcing, and on top of this place the remaining 2½ inches of concrete and tamp it down well.

This top surface will be the top of the finished bench seat; therefore it will pay to take pains to finish it to as smooth a surface as possible with a float and trowel. Cure the slab in the same manner as described for curing the pedestals. Do not attempt to remove the under part of the mold for at least 7 to 10 days; the side of the form can be removed any time after 48 hours.

After the forms are removed and the concrete has hardened, a good smooth surface can be given to the bench by wetting it down well and rubbing it with a fairly fine grade of carborundum brick.

When taking off the mold, watch to see if the cast is damaged in any way; the damaged parts can readily be replaced or filled in by applying and forming into shape cement mortar composed of 1 part cement to 1 or 2 parts sand. Before applying this mortar, be sure to wet down the surface of the cast thoroughly, since this is necessary to secure a good permanent bond between the patch and the original concrete.

In setting the bench up, place the pedestals about 7 inches in from the ends of the slab. It is unnecessary to secure the slab to the pedestal in any way, since its weight will keep it in place.

An interesting effect, which adds to the appearance of the ends, can be achieved by tooling the recessed panel in the outer sides of the pedestals. This is done by gently striking the surface with a chisel and hammer, thus giving a rough angular texture to that part of the pedestal.

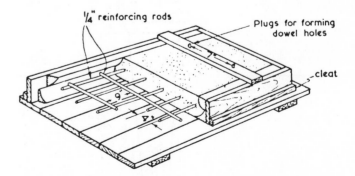

¼" reinforcing rods

Plugs for forming dowel holes

cleat

Fig. 13-4a *An alternate design for a bench slab using wood molding to form beveled edges. (Portland Cement Institute, Johannesburg, South Africa)*

Other attractive surface effects can be obtained on the pedestals by using selected aggregates or stones. The concrete should be mixed as previously described, except that instead of using pea gravel, you may use: marble chips, broken chips of red brick, or other coarse aggregate. The size of the pieces should not exceed ½ inch to ¾ inch. Place the mixture in the mold as previously explained, but instead of allowing it to remain there for 2 days as before, remove the forms after 18 to 24 hours. The concrete will be a little soft. By hosing down the concrete with water and gently brushing the surface with a good stiff brush the surface cement paste will be removed. The stone or marble or red brick will gradually be exposed, giving an exposed aggregate surface.

If any surface cement film should remain on the stones or exposed aggregate after the above treatment, apply a solution of 1 part commercial muriatic acid to 4 parts of water to the surface of the cast by brush. This solution should be allowed to remain on the surface for 15 to 20 minutes, and then thoroughly cleaned off with good clean water and a stiff brush. This acid treatment will cut away the cement film and leave all the stones clean and bright.

Concrete Pedestal and Sun Dial

A wide variety of bird baths and sundials are available as precast items from concrete products manufacturers, building centers, and garden supply centers. But if you decide to build your own, here is how to build a simple pedestal (Figure 13.5).

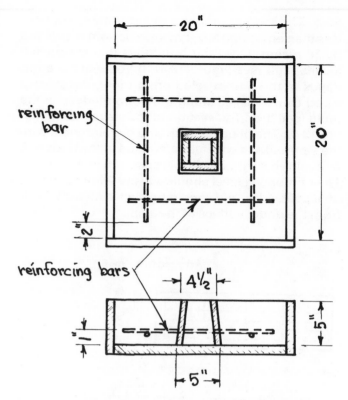

Fig. 13-6 *Details of base mold. It is a good idea to add reinforcement to the concrete base.*

The mold should be made of 1-inch lumber or plywood. The base mold (Figure 13.6) consists of just a square box with sides 5 inches high. In the center of the bottom of this box place a tapered core, to produce a hole in the cast that corresponds in size to the outside dimension of the plug on the bottom of the shaft of the pedestal (see Figure 13.5).

Fig. 13-7 *Details of cap mold. Wood molding is used to achieve a round flair to the cap.*

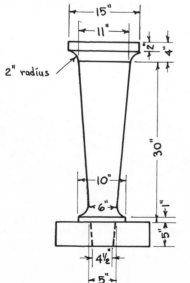

Fig. 13-5 *A sun dial pedestal composed of cap, shaft and base.*

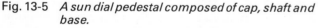

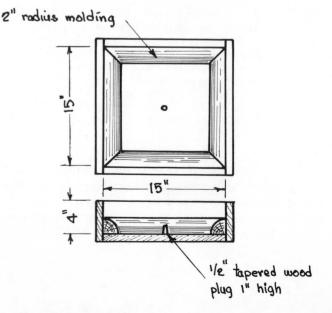

The mold for the top or cap of the pedestal is shown in Figure 13.7. This also is a square box, but it is 4 inches deep. Place a ½-inch tapered plug in the center of its bottom, for a ½-inch hole at the bottom of the cap so you can insert the ½-inch reinforcing rod which passes through the length of the shaft. Put strips of 2-inch, quarter-round stock molding mitered at the corners in the bottom of this mold to give the desired outline to the lower portion of the cap.

The main shaft mold is made in three pieces: A, B, and C (Figure 13.8). Figure 13.9 shows the details of the sides of the main part of the shaft mold (A). Figure 13.10 shows details for parts B and C. Part B is a bottomless box 10 inches square on the inside, with sides 6 inches high. Inside on all four sides, and mitered at the corners as shown, nail pieces of 2-inch quarter-round stock molding. These are securely fastened to the sides, 1 inch from the top. Part C of the mold is made of four pieces of 1-inch board (or plywood) on which is built up the cone which forms the lug on the bottom of the shaft. Part A of the mold at its top should have fastened to it, on all four sides, pieces of 2x1-inch tapered strips (as shown by shaded portion at "d" in Figure 13.8). The outside dimensions of these strips should be such that the inner portion of part B fits over them snugly.

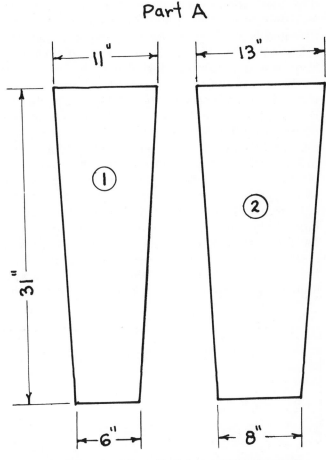

Fig. 13-9 *Main part (Part A) of the mold for the shaft. Make two each of piece #1 and piece #2.*

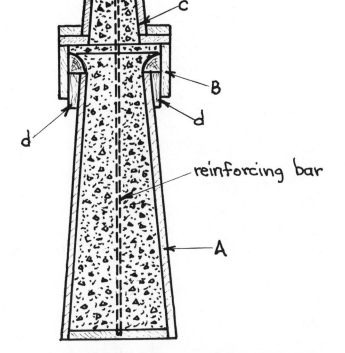

Fig. 13-8 *Assembled mold for casting the shaft. Assembly parts are noted by A, B, C; strips are indicated by D. A reinforcing bar runs through the middle.*

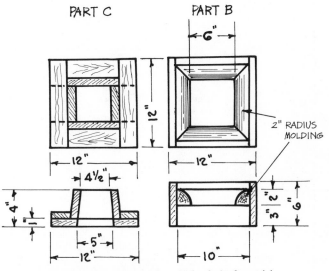

Fig. 13-10 *Details of Parts B and C of shaft mold.*

In the bottom of part A of the mold, bore a $9/16$-inch hole a half-inch deep in its center for insertion of the #4 (½ inch) steel reinforcing rod.

After completing the various parts of the mold described above, sandpaper their inner surfaces and give them two coats of shellac. Let this dry thoroughly and then oil the inside surface well with a fairly thin oil. This will aid in getting a smooth surface and in stripping the forms after casting.

Now assemble the shaft mold, letting section A stand on end as shown in Figure 13.8. Place section B in position as shown, care being taken to let the quarter-round molding rest snugly down on the pieces "d" of section A. Then place the reinforcing rod in position, and start to deposit the concrete.

A concrete mixture of 1 part cement to 2 parts sand to 2¼ parts pea gravel can be used. Or packaged mortar and concrete, available through building supply centers, can be used. If a white shaft is desired, use one part white portland cement and two parts white marble screenings, ranging in size from dust up to ⅜ inch. Mix these together dry, and then add enough water to make a fairly thick paste.

Fill the mold flush with the top of part B, tapping the sides of part A of the mold, and rodding or spading the fresh concrete occasionally to settle the concrete mixture as it is being placed.

When the concrete is flush with the top of section B, place section C in position, and proceed to fill it flush with the top.

Allow the concrete to set or harden in the mold for at least 24 hours before attempting to remove the molds.

In fastening the molds together use as few nails as possible. When removing the molds from the cast concrete, take care in loosening them so as not to injure the concrete. In removing the shaft mold, take off part C first, then part B, and finally section A.

The base and cap molds should be filled with the same concrete mixture as used in the shaft. The concrete should also be allowed to set for at least 24 hours before removing the forms. It is a good idea to insert into the base when casting, four pieces of #4

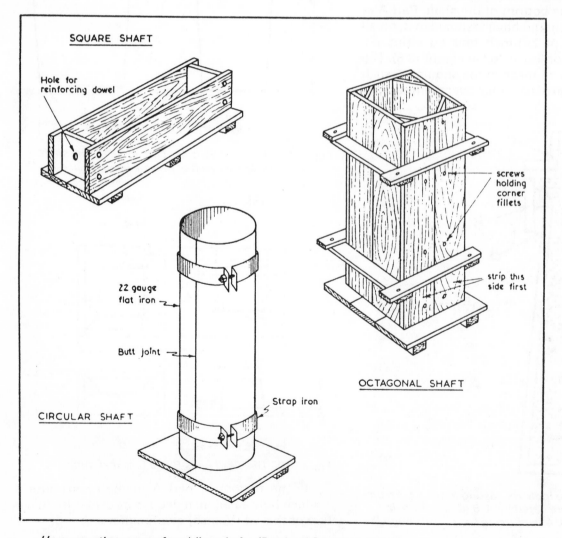

Here are other ways of molding shafts. (Portland Cement Institute, Johannesburg, South Africa)

(½ inch) reinforcing bars placed as shown in Figure 13.6. A square piece of welded wire fabric can also be used. The reinforcing adds strength and will prevent cracking in case the foundation upon which the pedestal is placed is not perfectly true and level.

If by any chance you damage the concrete while removing the molds, wet it down with water and point it up with a mortar similar to the original mix. (For the white mix described earlier, use a mortar made of 1 part white portland cement and 1 part of marble dust, mixed with enough water to produce a fairly thick paste.)

After pointing up the various parts of the pedestal, let the parts stand for a short time. Then all of the pieces should be kept moist for about 7 days—either soaking in water or covered with wet burlap.

After the pieces are hardened thoroughly they can be assembled or set up in position. They can be joined together with an epoxy adhesive or with cement mortar. If using cement mortar, the surfaces of the parts which are to be joined together should be sprinkled with water and covered with a thin layer of cement mortar. (For the mixtures described earlier use 1 part white portland cement and 1 part marble dust.) Place the parts on top of each other and work around with a twisting motion until they are bedded in place. Smooth off the surplus paste forced out at the joints. Allow the pedestal parts to set, without being disturbed, for 1 to 2 days, at which time they will be firmly secured in place.

A sun dial is usually made of brass or bronze. Once placed on the pedestal its weight would hold it down, but it is better to cement the sun dial to the pedestal with an epoxy adhesive (or a cement mortar). When placing a sun dial, always see that its vane points to the north and that the pedestal is placed in the full rays of the sun.

You should prepare a good solid foundation for the pedestal to rest on. Otherwise it may settle as the ground softens in springtime.

The size of the foundation should match the size of the pedestal base and be 2 to 3 feet deep. It is easy to build such a foundation: just dig a hole of the desired size and depth, and fill with concrete. Use prepackaged concrete, or a typical driveway mix, or even a mix as described earlier for the pedestal. Add enough water to make it a thick pasty mass. Tamp it down well, level it off, and allow it to set or harden for 24 hours, at which time the pedestal can be placed in position on it.

Fig. 13-11 *Pieces of wire mesh to form the frame for a square garden pot.*

Concrete Pottery and Planters

Another fun project is making concrete pottery. Anyone familiar with working in clay can model in concrete. But remember that portland cement mortar has to be worked differently. Cement mortar is a mixture of cement, sand or rock dust, and water; it cannot be modeled like clay but must be placed in a mold or form.

Making a Square Pot

Wire forms. There are several ways to make forms. One is to make wire frames on which to plaster the mortar, and another is to make wood or plaster molds. Use of wire forms is simplest, although wood or plaster molds should be built if you plan to make a quantity of units of the same shape. This combination of wire and mortar is a simplified version of what is called "ferro cement," multiple layers of mesh and mortar used in the construction of concrete ships.

The best material for making wire forms is galvanized wire lath having about a half-inch mesh. This can be found at almost any building supply center. The only tools needed are tinsnips and pliers. The frames can be shaped for either square or round pottery.

To make a wire form 5 inches square by 4 inches high, first cut with tinsnips a piece 5 inches on a side. Then a longer piece of wire lath is cut to desired height and length. The length will be the sum total of the sides plus 2 inches—in this case, 22 inches. This allows 1 inch for lap and ½ inch of surplus wire on each end, once the piece is bent into a square (see Figure 13.11). Since the height of the finished form is to be 4 inches, cut the lath to 4½ inches, leaving a series of wire strands half an inch long at the bottom. Then bend four square corners at 5-inch spacings, making sure the lap comes roughly in the

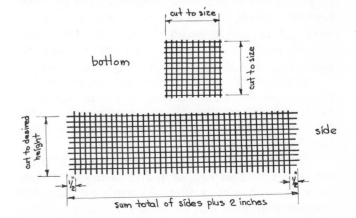

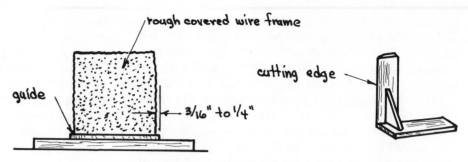

Fig. 13-12 and 13-13   *Left: The base "wood form" can be used as a guide in placing the finish mortar coat on the rough covered wire frame. Right: A template to help in keeping the sides of the planter plumb and even.*

center of one of the sides. The free ends of the wire are bent around the strands of the mesh and clinched tightly to secure the sides. Be sure the corner lines are perpendicular to the base. After completing the side frame, place the bottom in position, and wrap the half-inch lengths of wire left at the bottom of the sides around the bottom strands to secure in place.

The cylindrical frame can be made in a similar manner. Of course, any size or shape can be made up, even hexagons or octagons.

Plastering the wire frame. Next cover the wire form with cement mortar. First comes application of the rough or scratch coat. The mortar for the scratch coat is made of one part portland cement and two parts of a fine, clean sand—a 1-to-2 mixture. Mix the sand and cement thoroughly in a dry state. Then wet down with water and mix thoroughly. Be careful not to get the mix too wet, or it will not hang on the wire mesh. For proper consistency it should resemble a stiff paste.

Apply the mortar with a trowel, or putty knife, or even an old table knife. Take as much mortar as can conveniently be handled on the end of the knife. Begin at the bottom of the sides of the wire frame and force the mortar well in between the meshes. Continue this operation until the entire sides of the wire frame are covered. Then turn the frame bottom-side up, and cover the bottom mesh the same way. The rougher the surface, the better.

After completely covering the frame, let the mortar set (harden) so that it will be securely bonded to the wire mesh. The mortar will usually harden enough in 4 to 5 hours for the form to be handled without damage.

The finishing coat comes next. The mortar for the finishing coat can be made of gray or white cement, sand or rock dust (like marble dust); even colored pigments can be added. A variety of tex-

tures and color are possible. One mix that gives a fairly light surface, with "sparkle," is 1 part portland cement to 2 parts of marble dust. Mix to the consistency of a heavy paste, as before. In placing the finish coat, it is advisable to build guides or templates rather than depending on accurately eyeballing the surface.

For the square form, cut a wood base (half-inch lumber or plywood) about ⅜ or ½ inch larger than the square rough coat of the pot. Lay the pot bottom on this base, leaving 3/16 to ¼ inch projecting beyond the rough coat all around the square (Figure 13.12). The finish coat must be built out to the edge of that base.

The next step is to make a template or forming strip for the sides of the pot. Since the sides are straight, a straight piece of wood can be used. It should be at least 2 inches longer than the side of the pot and mounted on a simple stand, so that it will be perpendicular at all times. Bevel the cutting edge of the template. See Figure 13.13.

Now plaster on the finish coat. First rough up the scratch coat with a sharp tool; then wash and brush off any loose particles. Apply the finish coat just as you did the scratch coat, first at the bottom and building out to the edge of the base template.

Cover the whole surface with the finishing coat, gradually building it out to the required thickness. Now take the template and slide it along each side, cutting off the surplus mortar and giving a smooth, true surface to the sides.

At this point the top of the sides might be a little rough. This top edge can be smoothed by hand, or the template can again be used to help even off the top. A piece of wood can be nailed to the top of the upright template at the proper elevation for scraping the top level and smooth.

In finishing the inside, the rough surface should be scratched and washed as was the outside surface before plastering on the finish coat. Using the

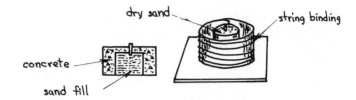

Fig. 13-15 *The completed mold wrapped in string and the core filled with sand.*

outside surface as a guide, it is easy to true up the inside with a thin straightedge or the knife used for plastering.

Remember to cure the pot. Protect it from the hot sun and keep it damp—with either a continuous spray of water or draped with wet burlap.

Once the concrete has been cured, the pot is ready for filling with a potting mixture and plants.

Casting Circular Pottery

Simple circular pottery can be cast with even fewer tools. The forming material can be cardboard, which has the advantage of easy stripping. Tin cans may be used, but they should be coated with a light oil or parting compound.

First make the outer mold (Figure 13.14) of heavy cardboard. Form this into a circle, lap the edges at least 2 inches, and secure the ends by glue or by sewing them together with heavy thread. Now cut out a circular piece of cardboard to fit the bottom of the cylinder. Sew the bottom to the outside mold, forming a circular container. Then make the core—that part which forms the inside of the vase. Select a core size that allows about 2 inches for concrete wall thickness. Form a cylinder, lapping and securing the edges the same as for the outside mold. Now, to hold the core in the center of the outside cylinder mold, secure small strips of heavy paper to the bottom of the outside of the core with glue (as shown). Then place the core in the bottom of the round box. Center it. Secure it in place by gluing down the small pieces of paper attached to the outside of the core.

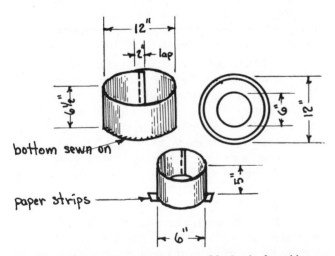

Fig. 13-14 *Round garden pottery: Method of making an outside mold and core out of heavy cardboard. (Molds can also be made from thin metal sheets.) The paper strips glued to the core are used to fasten the core in place on the bottom.*

Now fill the inside of the core with sand. This prevents the core from collapsing when the concrete is placed in the mold. Before placing the concrete, bind the outside mold with heavy twine (see Figure 13.15) to keep it from bulging. Now insert into the sand, in the center of the core, a tapered wood plug about ¾ inch in diameter. Oil or shellac it so that you can draw it out easily later. Let it project about 2 inches out from the core. This plug forms the drainage hole in the bottom of the vase, and the mold is now complete.

A typical mix can be composed of 2 parts good clean sand, not too coarse, and 1 part portland cement. Mix the sand and cement together thoroughly while still dry until a uniform color is obtained throughout. Now add enough water for a consistency like putty or fairly stiff dough. Place this mixture in the mold, ramming or tamping it down lightly as you put it in. Fill the mold flush to the sides and level off (about 2 inches above the core). Do not disturb the mold for at least 2 to 3 hours.

If you want a flush circular bottom on the vase, the job is now done and you should keep the vase damp—either by keeping it covered with wet burlap or by soaking it in a tub of water.

For a softer edge on the bottom—a rounded appearance—shape it before the concrete gets too hard. Wait 2 or 3 hours after casting so the concrete will be hard enough for removal of the outer mold. Then cut the sharp corners roughly with a strong knife, or a mason's trowel. You can stop here; or, for a uniform rounded corner, use a template to shape the vase (Figure 13.16). Naturally, make this template before casting the vase. An easy-to-make template would be constructed as follows. Take a piece of fairly heavy sheet tin and draw on it an exact outline of the bottom half of the finished vase. Now cut a piece of 1-inch thick wood, following the same shape. Nail the tin template to it with the tin extending out ¼ to ½ inch from the wood. Hold the bottom part of this template firmly to the working board and against the side of the concrete cast as shown. Gradually work it back and forth around the piece so the superfluous mortar, which is still in a soft

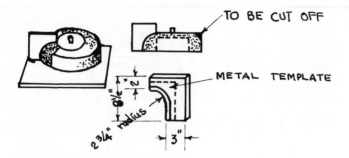

TO BE CUT OFF

METAL TEMPLATE

Fig. 13-16 *Shaping or rounding off the bottom of the concrete pot. A template with a cutting edge made from tin or aluminum can be used to cut and shape the edge.*

state, will be cut or scraped off the case and a good uniform outline will be produced around its entire surface.

If any holes or marked irregularities show up on the surface of the vase, point them or fill them up with a mixture similar to that used in the body of the vase. A good, smooth, fairly light finish can be procured by rubbing the whole surface down with coarse emery cloth. Then soak the vase in water to cure and to achieve a hard, durable surface.

You can follow these steps, adjusting for size variations, to make almost any size or shape of planter or vase.

Surface Finishes for Concrete Pottery

The ordinary concrete or mortar surface is usually a dull gray color. But if you want something a little brighter for your concrete planter or vase, you can incorporate color pigments in the mix or vary the aggregate.

In small work, where the thickness of the finished vase is about ½ inch, never use any aggregate exceeding ⅛ inch in size. In larger work having a thickness of 1 inch or more, aggregates up to ¼ inch can be used. Some interesting textures for pottery work can be obtained as follows.

A mixture composed of 1 part white marble chips, not exceeding ¼ inch in size, and 1 part of traprock or other dark stone of the same size, mixed with 1 part white portland cement, will produce a surface similar in appearance to a light granite. This mixture should set for 12 hours after casting. Then carefully remove the mold, as the concrete is still green, and brush lightly the surface of the concrete, using a stiff brush.

Because the concrete is not yet thoroughly set or hardened, this operation will remove the surface cement paste and expose the aggregates of marble

and traprock. After brushing, allow the piece to cure and harden for a few days. You can then treat the surface with a solution of 1 part commercial muriatic or hydrochloric acid to 3 parts of water. Dash this solution onto the face of the concrete surface with a brush and allow it to remain for at least 15 minutes. Then thoroughly scrub it off with a stiff brush and plenty of clean water for a surface full of life and sparkle.

To vary results, use white marble chips and crushed red brick, or various colored marbles crushed to the proper size. Then by treating the surfaces as explained above, the colors in the various aggregates will be exposed for an interesting surface.

To simulate white marble, use 1 part white portland cement to 2 parts of marble dust, and treat surface with acid as described.

To simulate red granite, use red granite chips or screenings. These can be secured at almost any stoneyard that cuts granite. The pieces should range in size from ¼ inch down to dust. If the pieces available are too large, they can be crushed with a hammer. The mix proportions should be 1 part portland cement to 2 parts granite. After the cast piece has set for 12 hours, brush the surface and treat with acid as already explained.

Mix proportions. It is always a good idea to mix a little more material than is needed, rather than not enough. When you start casting a piece of pottery you should continue until the mold is full.

Novices sometimes figure that if an amount of finished material equal in bulk to three cupfuls is required, all that is necessary for a 1 to 2 mixture, for example, is to take one cupful of cement and two cupfuls of sand, mix them together, and there will be enough material to fill three cups. This is not so.

The particles of cement are ground so fine that the cement is practically one dense mass. The particles of sand are coarser, and between each of the particles appears a space or a void. When the material is mixed, the cement fills in these voids. Therefore in that 1 to 2 mixture, the initial 3 cups will not give 3 cupfuls of finished material, but only about $2^1/_5$ to 2½ cupfuls. Keep this in mind as you mix your material so that you will have enough to complete your pottery.

When using larger stone aggregate in the mix, the spaces or voids between particles of stone are filled by the cement and sand, as the voids in the sand were filled by the cement. As in sand, the larger the particles of stone, the greater the percentage of voids; therefore, a greater amount of sand and cement are required to fill them.

It is always best to use a well-graded sand; that is, a sand in which the particles are not uniform in size. This means that the particles vary in size from about 1/32 inch or smaller to 1/16 inch or somewhat larger.

For these projects, be sure the sand is clean; that is, free from loam or clay. You can buy mason's sand at many building or landscape centers. You can recognize dirty sand by placing the sand in the palm of your hand and slightly wetting it. Then, if by rubbing it around the hand becomes discolored, there is dirt or loam in the sand. Dirt can be washed out by placing the sand in a pail of water and agitating it, causing the dirt to rise to the top.

When you start the casting, continue until the mold is full. Stopping in the middle can create a plane of weakness wherever the concreting was left off and started again. Also, do not use material that has been let stand for more than half an hour. In that time the concrete will have started to get its initial set. Do not remix the same mass; the concrete will not have the same strength as that of the freshly mixed material.

Garden Pools

A small pool with aquatic plants and colorful fish can become a quiet, restful corner in your backyard garden. It can also be inexpensive and easy to build. No special tools are required, beyond what every gardener already has, along with a few mason's tools.

A concrete garden pool watched over (left) by concrete toadstools. (Portland Cement Assn.)

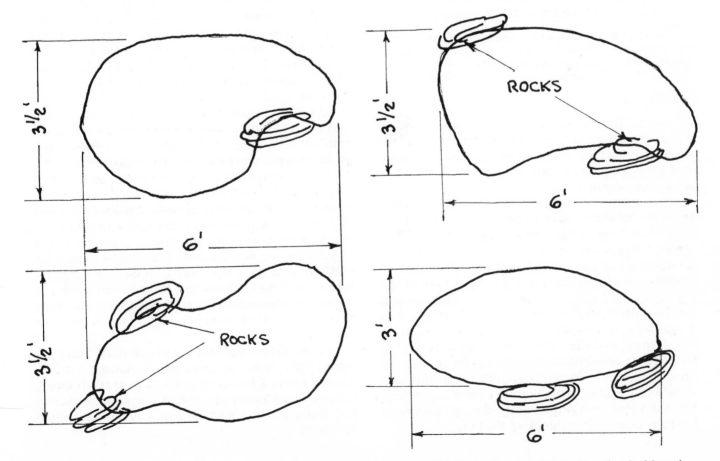

Garden pools can be built in free-form shapes and the edges lined with rocks or precast concrete slabs. Boulders can be added for accent.

The size and shape of the pond will depend on the layout of the garden and on the type of plants to be grown. Small species of water lilies will require a pool only 3 feet in diameter; larger varieties of lilies require an area at least 6 feet across. Pool depth is dictated to some extent by the type of plant. Water lilies require a depth of at least 22-24 inches. Allow for ledges or pockets to hold the shallow-growing plants.

A simple bowl-shaped or free-form shape is easily dug. Excavate about 3 feet deep in the center and taper the sides. To build a winterproof pool requires a good subbase. The pool should be built on compact ground with a 6-inch granular subbase of gravel, crushed stone, or slag.

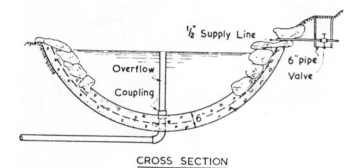

CROSS SECTION

Cross-section of a bowl-shaped or free-form pool. Rocks can be embedded in the concrete. Wire mesh reinforcement is placed in about the middle of the concrete. (Portland Cement Institute, Johannesburg, South Africa)

If the sides are not steeper than a rise of 1 foot in 2 feet, no forms will be needed. Floor and walls should be placed at the same time to avoid any construction joints. Start concreting at the bottom and work your way up the sides.

Place concrete about 6 inches thick. Smooth and finish the concrete with a wood float. A layer of reinforcement, bent to the shape of the pool, should be placed halfway into (in the middle) the concrete. The reinforcement can be 6x6—W1.4 x W1.4 (6x6—10x10) welded wire fabric or No. 3 (⅜ inch) reinforcing bars set 12 inches on center both directions. It can be placed on chairs to hold it in location, or a 3-inch layer of concrete placed first, then the reinforcement, and the concreting completed. This should be a continuous operation to prevent a cold joint between layers of concrete. A drain pipe should be installed in the bottom before concreting. Block the pipe with wood or rags during concreting to keep concrete from blocking the drain.

The concrete can be cured by covering with wet burlap. Or, even better, cure by filling the pool with water as soon as the concrete is hard enough to avoid marring or washing out.

If you want a cobblestone surface in the pool, instead of a smooth finish, the stones can be placed while the concrete is still plastic (see the section on cobblestone surfaces in Chapter 11). The pool can be edged with natural stone, a gravel bed, precast concrete patio slabs, or precast paving stones.

Although you can fill the pool with a garden hose, pipe connections hooked up to your house water system are desirable and make for a more permanent setup. The pool may be emptied by siphoning, but it is preferable to build in a drain by unscrewing the length of pipe above the coupling set flush with the floor. All plumbing connections should be made before concreting begins. When draining the pool, a wire cage or screen should be placed over the drain to keep debris or plants from blocking it.

Remember that the first water put into a pool will absorb the alkali from the new concrete. It is absolutely essential to fill the pool, and even to change the water several times, to permit the complete absorption of the free alkali into the water. This cleansing operation takes at least a month, during which the pond should be emptied and refilled two or three times. Each time you empty the pond, brush the sides down. Then drain and refill with fresh water before planting or stocking with fish. If you plan to stock the pool with valuable fish, it is well to test the pond water to verify that excess alkalies have been removed. Test the water by introducing a few small goldfish or tadpoles. If they show no ill effects, the pond is safe to stock with more valuable fish.

In northern climates, a well-reinforced pool will seldom crack over the winter. It is safer, however, to drain the pool than to try to cover it with boards and mulch or straw. If the pool is drained, and the plants left in, place a covering of manure, straw, or leaves directly on the soil bottom. The most satisfactory method is to have the lilies in tubs which can be taken out. The lily tubs can then be covered in a trench or brought into a cool basement and kept covered with moist wrappings of burlap (to prevent dry rot).

Water lilies require plenty of rich food. You may want to give your concrete pool a natural bottom. Spread about 2 inches of manure over the bottom, followed by 8 inches of sod-soil. Lilies will do well in this, but if the pool is small, you may not be entirely satisfied.

It is essential in a small pool to keep it well stocked with fish to prevent mosquito growth; fish will stir up a pond whose bottom is covered with soil, so that the water will be constantly muddy and not very attractive. To avoid this, the lilies can be planted nearer to the surface in boxes or tubs filled with manure and soil. The concrete bottom can then be left clear, and the surface of water covered with floating plants. In this way plants can also be easily removed when you want to clean the pool.

All Kinds of Things

There are so many things that can be built from concrete, some practical, some just for decoration, some simple in design, some complex.

Splash block may be one of the most practical and simple units available through garden supply centers and concrete products plants. Turf grids and precast patio slabs can also be used for the same purpose. Splash blocks are set under a downspout to carry rain water away from the foundation and to prevent erosion of the lawn.

Edging strips around flower beds can be straight, scalloped, circular. A variety of shapes can be purchased or you can also cast your own simple homemade forms.

Mailbox posts can be built of concrete block or brick: specially shaped precast units, or simple straightforward posts and boxes you can mold yourself.

All kinds of picnic tables are possible in concrete. Even end tables and benches come in a variety of shapes, and can be assembled from screen block or other masonry units.

Two styles of patio tables built with concrete screen block supports, and precast patio slabs serving as table tops. (National Concrete Masonry Assn.)

For this fire pit, 4x8x16-inch two-core concrete block were laid on side in a trench about 2½ inches deep. The weight of the second layer of block is enough to hold them in place. The open cores provide circulation of air for smoldering coals. The circular seating ring was built of the same units. (National Concrete Masonry Assn.)

Concrete block units painted in bright colors and stacked together form a simple patio bench. (National Concrete Masonry Assn.)

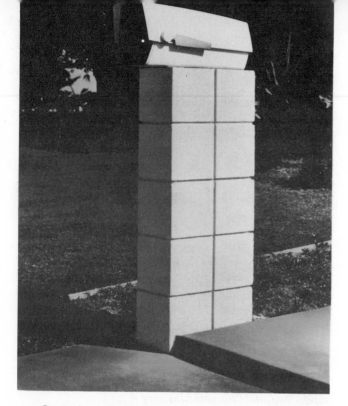

Square concrete block stacked to make a mailbox stand (National Concrete Masonry Assn.)

An exposed-aggregate concrete wastebasket. A plastic bag liner fits inside. The 22-inch square, 32-inch high container weighs 55 pounds. (Four Tomorrow, Inc.)

Castles in concrete are a modular system of cylindrical forms. A single cylinder is 2 feet 4 inches in diameter and 2 feet high. The double is 5 feet 6 inches long, 2 feet 4 inches wide, and 2 feet high. (Form Incorporated)

Playgrounds

While every home backyard is a playground, some of the best accessories can be built in concrete. Large precast pipe, painted with bright designs, can be laid flat to make ideal crawl-throughs. And some of the playground equipment available is, in all respects, also pleasing sculpture.

It is possible to cast a wide range of shapes in all kinds of textures and finishes: smooth for sliding, or rough for climbing. Concrete play sculptures are virtually indestructible. You can create concrete trees, castles, turtles, playwalls, and endless other shapes. Some units are precast; others can be built on the site.

Canoes and Toboggans

If you really want to try something different, build a concrete canoe or a concrete toboggan. For the past several years engineering students at universities in the United States and Canada have been building concrete canoes and toboggans and racing them in competitive meets. A wide variety of designs and combinations of materials have been used.

Concrete toboggans. Design and construction techniques can vary. Some type of reinforcing is required, but this can be bars, wire fabric, or expanded mesh. A typical design is a flat slab with a slightly curved and raised front end. The push bar at the rear can be steel reinforcing rods as can be the side hand holds. Dimensions are in the range of 8-10 feet long, 1½-1¾ feet wide, and about 2 inches thick. Weights have ranged from about 90 pounds to as high as 300 pounds, depending on design. The running surface can be cast completely flat and smooth, or with grooves and ridges aimed at improving ability of the "vehicle" to hold a staight course. It is even possible to build a saucer shape. Depending on the bottom surface—and snow conditions—the concrete riding surface may require waxing before the run.

Concrete canoes. Recorded evidence indicates that a Frenchman, Jean Louis Lambot, "invented" a small concrete boat in 1848 (which is on display in the Brignoles museum). It was built with a metal mesh plastered with mortar.

This system is commonly called ferro-cement: a combination of reinforcement mesh and a cement-sand mortar. The dense mortar is plastered onto a framework of mesh (which can be just plain chicken wire). This gives a strong but thin concrete slab or hull. Water-tightness is achieved by using a very

An all-concrete-canoe race. *(American Concrete Institute)*

A ferro-cement concrete canoe. *(American Concrete Institute)*

dense mortar mix. For a ¾ inch thickness, ferro-cement weighs about 10 pounds per square foot. It must be cured just like any other concrete; moist curing, preferably under wet burlap, should continue for 1 to 3 weeks.

What happens if your canoe hits a rock and punches a hole in the hull? The repair is easy and simple. Pound the damaged area with a hammer on one side, backing it up on the other side, to pulverize the mortar. Clean out the debris and particles and hammer the mesh back into shape. If the mesh was cut or ripped, new mesh can be woven into the gap. Then replaster the area with mortar. An epoxy bonding agent applied around the edges of the opening prior to plastering helps achieve good bond with the old concrete.

Concrete canoes raced at meets sponsored by the universities, American Concrete Institute, and American Society of Civil Engineers have proved many types of aggregates and reinforcement are feasible. Reinforcement has included: fiberglass strands, welded wire mesh, ¼-inch hail screen, fiberglass window screening, hardware cloth, and chicken wire. Aggregates have varied just as much: sand, polystyrene beads, expanded shale, volcanic ash, perlite, and styrofoam balls. Depending somewhat on overall size, aggregate, and reinforcement a 14-foot canoe can weigh anywhere form 65 to 400 pounds.

Concrete canoes can be molded over almost any form material. The very simplest form is to shape a pile of damp sand to the contours desired and to bend the wire mesh over that shape. More sophisticated and better forms include: (1) an open framework of mesh supported by wood bulkheads, (2) gypsum cement mortar molds, (3) plywood warped to contour shape, and (4) blocks of solid expanded polystyrene. The hull is "cast" upside down.

Unless you plan to build the canoe by halves and glue them together—and this, by the way, has been done—you want to plaster the canoe in one continuous placement. That means you want all materials, tools, and helpers on hand at one time.

As you might guess, with the wide variety of aggregates mentioned earlier, there is no one standard mix. A typical cement-sand mortar might be composed of 1 part cement to 2 parts of sand or 1 part cement to 2½ parts sand. Use an evenly graded mason's sand. Add only enough water to achieve a stiff but workable mix. The best way to mix the mortar is in a plasterer's paddle mixer.

Plaster the mortar on by hand and trowel. A small vibrator will help consolidate the mortar, but this can be accomplished by hand tamping and forcing mortar into the mesh with hands and trowel. Start at the bottom of the form (which will be the top of the canoe) and work up to the keel. Finish off the surface with a wood float and then trowel. Then cure for at least a week.

Once the canoe has hardened, it can be turned over and the inside smoothed off with a wire brush and sanded with a carborundum stone. Then sand the exterior and, when the surface has dried completely, you can paint it with an epoxy paint.

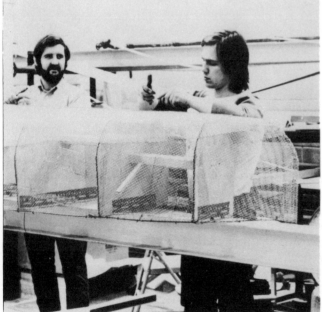

This canoe was formed by using bulkheads to hold the wire mesh which was bent to fit the contours. The concrete mortar was then plastered on and through the mesh. (American Concrete Institute)

The builders of this canoe laid the reinforcing mesh over shaped blocks of solid expanded polystyrene. The concrete is then plastered over and into the mesh. (American Concrete Institute)

14. Maintenance and Repair

Concrete Problems

Any material will deteriorate to some extent under weathering and over a period of time, and concrete is susceptible to both these influences. Accidents and wear-and-tear often leave their mark on concrete; freshly cast concrete can be damaged when forms are removed; and, we have yet to develop a formula for crack-free concrete. For all these reasons, a discussion of repair methods and substances will be given here.

There are all kinds of repair materials and techniques. Of those mentioned here, the emphasis will be on those which can be easily handled by the homeowner without special tools or equipment. For very complicated or extensive repairs, call in a contractor.

The table below lists various types of concrete damage with their suggested repair techniques and materials.

CONCRETE DAMAGE:
Repair Techniques and Repair Materials

Concrete Damage	Repair Technique	Repair Materials
Active cracks	Caulking	Elastic sealants
Dormant cracks	Caulking	Bituminous coatings
	Coatings	Elastic sealants
	Concrete replacement	Epoxies
		Expanding mortars
		High-speed setting materials
		Latex-modified concrete
		Portland cement concrete and mortar
Crazing	Coatings	Epoxies
		Latex-modified concrete
		Linseed oil
		Portland cement mortar
Dusting	Acid etching	Epoxies
	Coatings	Latex-modified concrete
		Linseed oil
		Surface hardeners
Efflorescence	Acid etching	Portland cement mortar
Small holes	Coatings	Dry pack
	Mortar replacement	Epoxies
		High-speed setting materials
		Latex-modified concrete
		Portland cement mortar
Large holes	Coatings	Epoxies
	Concrete replacement	Expanding mortars
		High-speed setting materials
		Latex-modified concrete
		Portland cement concrete and mortar
Popouts	Coatings	Bituminous coatings
		Epoxies
		Latex-modified concrete
Scaling	Coatings	Epoxies
		Latex-modified concrete
		Linseed oil
Spalling	Coatings	Epoxies
	Concrete replacement	Expanding mortars
		High-speed setting materials
	Mortar replacement	Latex-modified concrete
		Linseed oil
		Portland cement concrete and mortar

Adapted from "Concrete Repair Problems: Causes and Cures," *Concrete Construction,* November, 1969.

Identification Tips

Here are some definitions and explanations to help you recognize the more common types of concrete damage or deterioration.

Cracks. In most cases, cracks should be considered "active," that is, they continue to develop. An unstable subbase can result in uneven settlement and thus cracking. Improper jointing to handle temperature-change effects will result in active cracks.

A "dormant" crack means that it was caused by a factor not expected to reoccur, such as temporary overloading—for example, a car or truck driving over a slab that had not been built for that kind of a load.

Crazing. Crazing cracks are shallow cracks that form a hexagonal pattern. These cracks usually occur while concrete is still plastic, or shortly after the concrete has hardened. Crazing can be caused by the concrete slab drying out too rapidly—by a rapid loss of moisture from the surface of fresh concrete or by the concrete being placed on a dry subgrade. Other causes could be: too much water in the mix, or excessive finishing. Crazing cracks are usually dormant.

Dusting. Dusting occurs when the surface of the concrete becomes soft and rubs off readily under traffic. Common causes of dusting include: too-wet concrete mixes, excessive finishing, or inadequate curing.

Efflorescence. Efflorescence is the appearance of crystalline salts on a concrete surface. This is caused by water that migrates from the interior of the concrete to the surface as the water evaporates salts are deposited. It does not hurt the concrete, but mars its appearance.

Holes. Whether large or small, if a hole is not cleaned and shaped properly before patching, the patch will probably not hold. It is important to repair holes as soon as possible. Sometimes a "hole" is caused when concrete sticks to the form as it is being stripped (removed). This usually occurs when the forms were not properly coated with oil before concreting. Small holes, called bugholes, may occur along the surface next to a form and are due primarily to entrapped air bubbles. Honeycombing is another so-called hole, but of a different nature. Honeycombing results from the coarse aggregate being placed with an insufficient amount of mortar, because the mix is undersanded, or because poor placing techniques are followed. To repair these areas you must force mortar into the voids, or remove all loose and poorly bonded material and then replace it all with concrete.

Popouts. Popouts, shallow surface holes, usually occur in slabs. They are caused by expansion of a particle near the concrete surface. Wet or frozen shales, cherts, lignites, and limestones are likely to cause popouts. Some absorbent aggregates that expand when exposed to freezing also cause popouts.

Scaling. Scaling is the sloughing off of thin surface layers of concrete. Scaling can be caused by freezing and thawing of the concrete, use of de-icing salts on concrete which is not air-entrained, poor finishing practices, repeated wetting and drying of the concrete, or chemical attack on the concrete.

Spalling. Spalling is a loosely used term; it usually refers to chunks of concrete that have been broken from the surface by mechanical damage or impact. Spalling is also caused by corrosion of the reinforcing steel.

Repairs

Techniques

Most manufacturers furnish specific instructions on the use of their repair products...be sure to read and to follow them. Listed below are some basic techniques.

Acid etching. Safety precautions must be followed when working with acids. Most acids used for etching concrete cause burns when they come in contact with the skin, and some acids also give off noxious fumes. You should wear protective clothing, gloves, boots, and safety goggles.

Efflorescence can be removed with a 10 percent solution of hydrochloric acid. Sometimes acid etching is used to remove a material from the surface that might impair the bond of a patching material. Usually the concrete is brushed vigorously with a stiff broom with the acid solution. After etching, the acid solution should be thoroughly flushed from the surface as soon as it has ceased foaming.

Caulking. Caulking involves filling fairly narrow openings (cracks) with a plastic compound. Cracks can often be sealed with an elastomeric caulking material.

Coatings. Coatings are materials of liquid or plastic consistency that can be applied directly over concrete. Some coating materials are epoxy resins, bituminous compounds, linseed oil, and silicone preparations.

Concrete replacement. Sometimes conditions are so bad that only complete removal and replacement of the damaged concrete will solve the problem.

Mortar replacement. Mortar replacement is usually confined to shallow holes. The following steps should be followed to achieve a successful repair:

(1) Thoroughly clean and shape the cavity.
(2) Obtain a good bond between mortar and the old concrete.
(3) Vary the consistency of the mortar depending on whether the hole is in a floor or wall.
(4) Eliminate or reduce shrinkage.
(5) Cure thoroughly.

Sack-rubbing. Sack-rubbing will often improve the appearance of a concrete surface with stains or small bugholes. First spray the concrete with water. Then rub damp mortar over the surface and into the voids with a rubber float or a piece of burlap. Add enough white cement to the mortar to match the color of the surrounding concrete. And, of course, cure the concrete as usual.

A sand finish can be achieved the same way, except that you would rub a creamy, rather than a stiff, mortar over the surface.

Sack-rubbing is most effective when done shortly after forms are stripped. Any large voids should be repaired before sack-rubbing.

Selection of Materials

Pick the material to suit the damage and its repair technique. A wide variety of products are available. In all cases, follow the manufacturer's instructions.

Bituminous coatings. Bituminous coatings have the ability to resist water passage; this is their major advantage. They can be applied as thin coatings, and are often used to waterproof the exteriors of basement walls. However, they are messy and may deteriorate if fuel is spilled on them.

Elastic sealants. Many materials are used to seal active cracks. Sealants are either hot-applied or cold-applied.

Epoxy-based compounds. Epoxy-based compounds are used for many types of concrete repairs. Epoxy resins, when mixed with hardening agents, develop excellent strength and adhesive properties; they harden rapidly and resist water penetration. Setting time for epoxy-based materials is sensitive to atmospheric temperatures. Epoxies can be mixed with fine aggregates to reduce unit cost and make the repair material go further. Epoxy-based compounds are useful under three conditions: where an adhesive is needed to bond plastic concrete to hardened concrete or to bond rigid materials to each other; for patching; when a thin, strong coating is needed over concrete.

Expanding mortars and concretes. There are several kinds of products on the market, and they were developed to overcome or minimize shrinkage. This can be especially useful in patching.

One method uses a small amount of aluminum powder which, by chemical reaction, produces hydrogen gas that expands the mortar to fill the confined space. Another type of expanding mortar contains very fine iron aggregates and a catalyst. The iron rusts, causing the mortar to expand and fill the void into which it has been placed. The third expanding mortar is a cement that chemically prevents shrinkage.

High-speed setting materials. High-speed setting materials harden and develop strength in a matter of minutes. These products are available either as admixtures to be added to the mortar or concrete, or as ready-to-use materials requiring only the addition of water. Remember—if the material sets in 5 minutes or less, only a little can be mixed at one time and it must be placed quickly. Read and then carefully follow the manufacturer's instructions.

Latex-modified concrete. Latex-modified concrete is often used for resurfacing repairs because it offers good adhesive properties combined with high compressive and tensile strength. It is also flexible, has low water absorption, and weathers well. This type of concrete can be feather-edged when placing the repair.

Linseed oil. Linseed oil is used when scaling has occurred but without severe enough damage to warrant more extensive repair. The usual solution is a mixture of 50 percent linseed oil and 50 percent mineral or petroleum spirits (by volume). Linseed oil is particularly effective in protecting new concrete when applied before the first freeze and prior to the application of de-icing salt. Linseed oil penetrates the surface of the concrete to a depth of about

⅛ inch. The oil inhibits further damage by forming a film that water and salt solutions do not readily penetrate. When used on driveways, it should be reapplied periodically, about every two to four years. Linseed oil does not affect the skid resistances of concrete drives.

Portland cement concrete. Portland cement concrete is a common replacement repair material. It is usually used for fairly large repair areas.

Portland cement mortar. Mortar is used to repair holes as well as to resurface pitted concrete surfaces. Thickness for portland cement mortar repairs should be at least 1½ inches. The mortar is easily placed by hand. It is particularly important to compact the mortar thoroughly, especially when stiff mortar is placed in holes.

Surface hardeners. Surface hardeners form hard crystals that retard dusting on concrete floors or garage slabs. These porducts have a fluosilicate as a base. This chemical reacts with the calcium hydroxide and calcium carbonate found in dusting concrete floors. Surface hardeners are inexpensive

and easy to apply, but should be reapplied periodically.

Preparing the Surface to be Patched

Whatever material is used to repair a damaged concrete surface, the patch will only be as strong as the surface to which it is bonded. The surface should be clean and sound. It must not be contaminated with oil, grease, paint, or mud. The surface can be scrubbed clean with a water-soluble detergent. Heavy deposits of material should be scraped off before scrubbing. Also, there is no point patching over unsound material. Any scaling, crumbling, or loose material must be removed down to clean, hard concrete. Hand picks or chisels can be used to remove unsound material. The depth to which material should be removed from a damaged area will depend on the patching technique and patching compound chosen. If using portland cement concrete or mortar, the area to be repaired should be removed to a depth of at least 1½ inches. If a latex-modified mortar is used, the cut can be less since this com-

The steps in repairing a damaged concrete slab. Top row (left to right): Preparing the surface by cleaning out all loose debris. Priming the surface. Next, a grout coating is placed. The grout of portland cement and water should have the consistency of thick paint. Bottom row (left to right): Placing and spreading mortar should follow immediately. Finishing the patch with a trowel. Repair completed. (Reproduced from Concrete Repair, Concrete Construction Publications Inc.)

pound works well for thinner patches. For most patches, the edges of the hole should be cut roughly square, or even slightly undercut. After chipping out the unsound material, be sure to remove all traces of loose debris and dust.

Portland Cement Patches

When using portland cement concrete or mortar as the patching material, the first step is to soak the cleaned bonding surface with water for at least an hour, and preferably overnight, before patching begins.

Right and wrong method for patching a small shallow hole in a concrete wall or slab. (a) Wrong—This patch is too shallow. The featheredge will soon chip out, and this patch will not hold. (b) Right—The edges should be vertical. The patch is deep enough for a good body of mortar to be placed; this patch will stay.

If the patch is being made in new concrete, work should begin just as soon as possible after removing (stripping) the forms or mold. If the patch is in old concrete, you want to wait after soaking until all surface water has disappeared. This depends on weather conditions and porosity of the base concrete. The old concrete, just prior to placing the patching mix, should be damp but still slightly absorbent.

Next, a bonding layer is applied to the clean, damp surface. On horizontal surfaces use a grout of portland cement and water mixed to the consistency of thick paint; it should be forced into the base by firm brushing with a semi-stiff bristled brush. On vertical surfaces the bonding layer should be composed of 1 part portland cement and 1 part sand. This mortar should be mixed ½ to 3 hours before use and should be of plastering consistency when applied. Occasional mixing during this period will keep the mortar from stiffening, but do not retemper with water. The bonding layer should be applied to a thickness of about ¼ inch using a stucco brush, but do not apply it too far in advance of the main repair. Otherwise the bonding mortar will dry out.

In general, the mortar or concrete used in the patch should be of the same materials and in the same proportions as that of the base concrete. Since a patch tends to be darker than the surrounding concrete, it is a good idea to substitute white cement for a part of the ordinary gray portland cement used.

For repairing the normal shallow spall, the patching mortar should be built up in layers about ⅜ inches thick. Keep each layer moist for a day or two before placing the next, cross-scratching it to provide a good bonding surface for the next layer. If the patch to be filled is deep, it is often more practical to build a form over the area and to pack concrete behind it.

The patching mix should contain just enough water to give the mix an earth-dry consistency, so that when a pat of mix is squeezed it will "cake", just leaving a trace of moisture in the palm.

On horizontal surfaces, patching concrete should be vigorously hand-tamped into place. On vertical surfaces, the concrete should be carefully rodded into place, making sure that the concrete is well compacted and fills corners fully. Then trowel or float the surfaces to the desired finish.

The patch must be thoroughly cured. The patched area should be kept constantly moist for several days and, if practical, curing should continue for as long as a week. If the patch is not carefully cured, it may dry out and shrink away from the old surface.

Epoxy Bonding Agents

An area patched with an epoxy-based adhesive is more likely to achieve a better bond than a patch based on a portland cement mortar coating.

When using epoxy adhesives, follow the manufacturer's instructions. The epoxy materials usually come in the form of a two-component system —the base resin and a hardener. Once the two are mixed, the curing process begins with a highly reactive chemical process.

The quantity of bonding agent to be mixed at one time should not be more than can be used up within the pot-life of the agent. This period will be noted by the manufacturer, but generally pot life is about 2 or 3 hours. Pot life means the period during which application is possible. Most bonding agents remain tacky for 1-2 hours after the specified pot life. The new concrete for a patch should be placed anytime during this tacky stage. Tacky set usually begins about ½ hour after application of the epoxy agent, when the free solvents in the mixture have evaporated.

The epoxy bonding material is best applied by stiff bristle brush. Brushing assures that the epoxy fills all angles and pores of the surface. Then the stiff mortar is placed and finished in the usual manner.

When working with epoxies, adequate ventilation is necessary since vapors can irritate the eyes, throat, and lungs. Also important, keep the materials off the skin, since this may lead to severe rashes; always wear gloves. If you spill resin on the skin, wash it off immediately with lots of ordinary soap and water. Do not use a solvent; it will only cause greater skin penetration.

Latex-Modified Cement Mortars

Another repair material is latex-modified cement mortar. For small repairs, the cement and aggregate are added to a latex emulsion already in a container. A latex-modified mortar is quite fluid. It also sets up rapidly. Manufacturer's instructions should be followed closely.

Mix latex mortars for 2 to 4 minutes until they reach the consistency of a smooth paste. Latex-modified mortars should be placed and finished immediately. In most cases they should be cured, preferably with wet burlap, immediately after finishing. (This is not always so, however. Some manufacturers do not recommend any curing, so follow that particular manufacturer's instructions.) With the rapid setup time of the patching compound, operations must be completed with minimum delay.

Before patching, the concrete should be roughened, cleaned, and presoaked. Best results are obtained by first brushing a portion of the mix into the pores and crevices of the bare concrete. Then place the remainder of the mix (which by this time may have started to stiffen) up to the desired finishing level. On vertical surfaces a two-layer approach helps reduce the surface skinning and permits a smoother job; on horizontal surfaces this is not necessary.

Crack Repair

Crack repair depends on how extensive the crack is, its size and depth, and whether it is an active or dormant crack.

If the crack is active, an elastic sealer or caulking material is the best way to seal the crack and still allow movement.

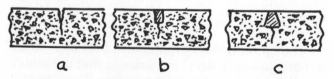

Repairing a wide crack (a), about ¹/₁₆ inch or wider. There are two good ways to cut out the crack for patching: (b) a vertical cut for the edges or (c) an undercut.

The dormant crack can be filled permanently with a patching mortar or compound: portland cement mortar, epoxy mortar, or latex-modified mortar. For a hairline crack, sometimes a grout made of portland cement and water is sufficient. The grout is mixed to a thick paste consistency. The paste is forced into the crack with a trowel or putty knife, and then smoothed level with the concrete surface. As in all repairs, the crack must be cleaned before filling, and the surface should be dampened prior to using the portland cement grout. This kind of patch should be moist cured.

If the crack is more than ¹/₁₆ inch wide it should be chipped away and widened with a hardened-steel chisel. To hold the patching material, the sides of the cut should either be straight or undercut. For small repairs, a prepackaged mortar mix is ideal. These come in packages as small as 5 pounds. A latex-modified mortar or epoxy mortar works extremely well in these situations. If you use a straight portland cement mortar, a typical mix would be q part cement to 3 parts sand. Follow the manufacturer's instructions in application. Most repairs call for the old concrete to be well soaked with water and damp at the surface when the mortar patch is applied, and it does not hurt at all to soak the repair area overnight. Once the area has been cleaned thoroughly, force the patching material into the crack, then smooth, and cure.

Repairing Steps and Curbs

Although you might be able to live with a cracked or spalled area in a sidewalk for a while, safety soon becomes a factor. Damaged curbs and steps are more than just unsightly; they are dangerous. Broken corners can be patched with portland cement mortar, but some of today's ready-to-use patching mortars are even easier. Latex-modified, acrylic-modified, and epoxy-sand-cement mortars are all effective.

The broken corner should be chiseled and cleaned out. Undercut the edges. Simple formwork may be required to hold the patch if the area is extensive. At other times the stiff patching mortar will support itself. The patch is tamped and compacted, then floated and troweled. Portland cement mortar requires curing—keep the concrete damp —as do some of the compounds. Wet burlap is a good curing method.

By the way, if the broken corner or piece is still in good shape, it is possible to glue it back in place. Clean the area and use an epoxy or latex mortar (some products are premixed, some you will have to

Steps in repairing a broken concrete corner—applicable to steps and curbs—with a prepared patching mortar, Thorite.

a. Cut out all disintegrated concrete until a solid surface is reached.

b. Brush out all loose concrete, chips, and dirt. Dampen area with clean water.

c. While area is still damp, apply a grout coat of patching compound. Work the grout well into the area to be patched.

d. Mix patching compound to a mortar consistency.

e. Force the patching mortar into the area to be repaired, in layers not exceeding 1 inch.

f. Scratch first layer while material is still soft, to assist in maintaining bond between layers.

g. *Apply next layer of patching mortar.*

h. *Overfill the area so that the patch is slightly larger than the surrounding area.*

i. *Shave off excess until patch conforms to the adjacent area.*

j. *The patch has been completed and repainted. (Standard Dry Wall Products.)*

mix). Butter the broken piece with the patching mortar and hold or brace it in place, usually 10 to 15 minutes. After the mortar has stiffened, the excess that has squeezed out can be cleaned off with a trowel or putty knife.

Concrete Stains

Staining may be due to natural causes, such as run-off water depositing soot or dissolved mineral salts. And, while you may keep the family automobile in top condition, invariably there is an occasional oil leakage on a garage floor. Or someone spills a can of paint. Fortunately you do not have to be either a chemist or a magician to remove most

stains. Almost any stain can be removed from recently cured concrete; old, long-neglected stains may require repeated treatments.

To preface a description of specific treatments, here is some general advice on procedures and precautions. The materials and chemicals mentioned below are readily available at drugstores, chemical or laundry supply houses, and possibly at paint or building supply centers. Under normal circumstances the cleaning materials can be used both indoors and outdoors without danger. Naturally, two very basic elementary safety precautions should be taken: (a) wash hands thoroughly after use, and (b) maintain adequate ventilation.

After the stain has been removed it is good practice to wash the area thoroughly with clean water to remove any residual contamination or particles of filler, and to be certain that no soluble and possibly detrimental salts remain on the concrete.

If you are not exactly sure what the stain is, it sometimes is better to test a cleaning treatment. Improper materials or techniques could result in spreading a stain over a larger area than originally involved. The best method is to prepare a small trial quantity of the cleaning agent and to apply it at the most inconspicuous point to assess its value. The composition or strength can then be varied appropriately. This trial-and-error approach applies particularly to the fillers used to form a paste. Different fillers have varying abilities to cling to vertical surfaces.

Also remember that on old concrete, accumulated dirt also disappears with the stains. A limited clean area in a sea of darkened concrete may, in effect, seem just like another stain. Time, wear, and weathering should solve this problem, so that the spot will blend back in with the rest of the concrete.

Treatments

Rust stains. Rust stains on concrete are common. They usually result from weathering of steel or iron attached to or resting on the concrete.

The cleaning materials needed are: socium citrate crystals, crystals of sodium hydrosulfite, and a paste of whiting and water.

The surface should be soaked with a solution made of 1 part sodium citrate crystals in 6 parts of water. Dip white cloth in this solution and paste it over the stain for 10 to 15 minutes. Or, apply the solution by brush at 5- to 10-minute intervals until the area is thoroughly soaked.

On horizontal surfaces you can sprinkle a thin layer of sodium hydrosulfite crystals, moisten with water, and cover it with a paste of whiting and water.

On vertical surfaces: soak with the sodium citrate solution, place the paste of whiting on a trowel, sprinkle it with sodium hydrosulfite crystals, and plaster the paste over the stained area—making sure the crystals are in contact with the stained area.

Allow the paste to soak in for at least 10 to 20 minutes. Do not leave the paste in place more than one hour, as black staining may result. Remove and flush with clear water. If stain remains, repeat the treatment, but usually a single application is adequate.

Aluminum stains. Aluminum stains usually show up as white deposits on the concrete. The cleaning agent required is a 10 to 20 percent muriatic acid solution. (Remember to observe the label precautions, since muriatic acid can affect eyes, skin and breathing.)

The white deposit may be removed by scrubbing with the 10 to 20 percent muriatic acid solution. On colored concrete, weaker solutions should be used. Flush with clear water after removal to prevent etching.

Grease stains. Grease normally will not penetrate very far into good dense concrete. Scrape off all excess grease from the surface and scrub with scouring powder, soap, trisodium phosphate, or detergent.

If staining persists, try a solvent. Avoid using free solvents such as gasoline or kerosene, since these only increase the degree of penetration.

Make a paste using a solvent and inert powdered filler. The solvent can be benzene, refined naptha solvent, or a chlorinated hydrocarbon solvent such as trichloroethylene. The filler can be hydrated lime, whiting, or talc. Apply the paste to the stain and do not remove until the paste is thoroughly dry. Repeat the application as often as necessary. Then scrub with strong soap, scouring powder, trisodium phosphate, or detergent (some are specially formulated for use on concrete). Rinse with clear water at end of treatment.

Oil stains. Lubricating or petroleum oil readily penetrates into the concrete surface. With any oil spillage there will be little danger of staining if the free oil is removed promptly. It should be soaked up immediately with an absorbent material such as paper towels or cloth. Wiping should be avoided, as it spreads the stains and drives the oil into the concrete.

Cover the spot with a dry, powdered, absorbent inert material (hydrated lime, whiting, powdered talc or portland cement). Leave it for one day. Repeat this treatment until no more oil is absorbed by the powder. If a stain persists or if oil has been allowed to remain for some time and has penetrated the concrete, other materials will be necessary.

Oils that have solidified should be scraped off as much as possible. Then scrub the area with a clean strong soap, scouring powder, trisodium phosphate or proprietary detergents specially formulated for use on concrete. By the way, a water softener from the laundry room—Calgon—works extremely well. Wet down the concrete surface and sprinkle on the Calgon (or similar material). Let stand for a little while. Then scrub and flush with water. This will remove most of the free oil. The following treatments can also be used (they are somewhat similar):

1. Make a paste of a suitable solvent, such as benzol, and an inert powdered filler (hydrated lime, whiting, or talc). Apply the paste to the area and allow it to remain in position for at least 1 hour after all solvent has evaporated. Remove and scrub with clear water. Repeat as necessary.

2. Make a poultice with a solution of 5 percent sodium hydroxide (caustic soda). Let dry for 20 to 24 hours, remove, and scrub the surface with clear water. Repeat as necessary.

Paint stains. Dried paint films can be removed satisfactorily by most commercial paint removers. Probably the most effective paint removers are based on methylene dichloride, and are available as liquids, gels, or pastes. Do not use paint strippers that contain acetic acid; these will remove the paint, but may also etch the concrete surface.

The remover should be applied liberally to the area and allowed to penetrate the film for 20 to 30 minutes. Gentle scrubbing will then loosen the paint film and allow it to be peeled or washed off. Wash with water. Any remaining residue can be scrubbed off with scouring powder. Color that has penetrated the surface can be washed out with dilute hydrochloric or phosphoric acid.

This treatment can be applied also to dried enamel, lacquer, or linseed-oil-based varnish. For shellac stains, the paint remover is replaced by alcohol.

Paint removers should not be used on freshly spilled paint or on films less than three days old, since they only tend to increase penetration of the fresh paint into the surface. Absorption with soft cloth or paper towels, followed by vigorous scrubbing, is recommended.

Coffee stains. Coffee stains can be removed by applying cloth saturated in glycerin diluted with four times its volume of water.

Soot. Soot and smoke can be removed by scrubbing the area with ordinary scouring powder. This is followed by an application of sodium hypochlorite (ordinary bleach).

15. Appendices

A. De-Icing in Winter

The best protection against severe winter exposure is good concrete. In northern climates the use of air-entrained concrete is a must. Concrete should contain about 6 percent air. Other measures to assure quality concrete include: a low water-cement ratio, care in finishing, and adequate curing.

If possible, sidewalks and driveways should be constructed well in advance of winter. Placed in the summer, they have plenty of time in which to develop strength before the freezing and thawing cycles start. Ideally, de-icers should not be used for removing ice and snow during the first year a sidewalk or driveway is down. Realistically, of course, this sometimes cannot be avoided if icy sidewalks become a safety hazard. However, wait at least 6 weeks after the concrete has been cured before applying de-icers.

Of the several chemicals available for removing snow from sidewalks, and driveways, and steps, the two best are sodium chloride (rock salt) and calcium chloride. They can be purchased at building supply centers, hardware stores, and even supermarkets.

There are some chemicals that are extremely corrosive to even the best quality concrete. Do not use them! Ammonium sulfate or ammonium nitrate will melt ice, but they also attack concrete.

De-icer chemicals work this way...When exposed to moisture, sodium chloride or calcium chloride generates heat and goes into solution, forming a brine that melts ice and loosens the bond between ice and the paved surface.

All de-icers are impractical where the snowfall amounts to more than 2 inches. You must plow or shovel first, then use a de-icer to melt the thin layer of ice or hardened snow that remains.

How and where you apply the de-icer may depend on the type of snow that is falling or forecast.

a. Dry, powdery snow can often be shoveled or plowed away, leaving a bare surface.

b. Sleet or freezing rain must be melted off with chemicals.

c. Wet or heavy snow can be melted off. Apply the chemical as soon as this type of snow begins falling to keep it from bonding to the paved surface. If it accumulates more than 2 inches, of course, it needs to be shoveled away.

Any de-icer will attack poor quality concrete or non-air-entrained concrete. However, quite often such concrete can be made more resistant with a protective surface coating of linseed oil or other material.

You can spread calcium chloride or sodium chloride at the rate of about ¼ pound per square yard. You can use a shovel, scoop, empty coffee can, or a small garden-type spreader. Do not overspread. Use only about a cup full for each square yard. Spread the material evenly; there should be no piles.

B. Hot Weather Concreting

You cannot always do your concrete work on cool, pleasant days. Summertime construction usually has to contend with hot weather.

A day that is very warm and breezy, with extremely low humidity, is a perfect day for sailing...but it can spell trouble for concrete work. However, by following tried and proven rules, you can get excellent concrete despite the temperature.

When temperatures are high (in the 80's) and humidity is low and with a brisk breeze, you can expect excessive and rapid evaporation of moisture, and the exposed surface will dry faster than the rest of the concrete. The ideal setting temperature for concrete is 73 degrees Fahrenheit. As the heat increases, however, setting time accelerates. If the temperature gets above 80 degrees there is a great danger of a "quick set" and of permanent damage to the strength of the concrete.

Other problems can result from excessive heat in the mix, which can occur if the mixer has to run too long. Heat builds up as the mixer drum turns, and the concrete becomes stiffer. To restore workability of the concrete it is frequently necessary to add water, *but this reduces the strength of the mix* and should be avoided.

The most important thing to remember when placing concrete during hot weather is that too-rapid evaporation is concrete's worst enemy. The concrete you mix yourself and that which comes in a transit-mix truck are mixed with specific quantities of water. If the subbase, the aggregates, and the forms you use are dry, they will absorb the water and cause rapid drying or hardening.

Everything should be well moistened before placing concrete. Thoroughly soak the subgrade the night before and sprinkle it again just before concrete is placed. Otherwise plastic-shrinkage cracks will occur during finishing.

Have plenty of help available to handle the concrete when it arrives; this avoids costly and troublesome delays. Discharge concrete immediately from the ready-mixed truck.

After the concrete is placed, every precaution should be taken to prevent evaporation of water. In severe cases windbreaks are erected to keep the sun, hot drying winds, or low humidity from drying the surface too quickly. Covering the surface immediately after finishing with tarpaulin, burlap, straw, membrane curing compound, or any other moisture-retaining material will prevent evaporation, permit the concrete to cure properly, and give stronger more durable concrete.

For the best results in hot weather remember these important steps:

1. Prepare the forms properly.

2. Have plenty of help and the right equipment for concreting.

3. Dampen the subgrade and forms.

4. Do not place more concrete than can be finished rapidly. Place the concrete as soon as delivered.

5. Protect fresh concrete from sun and wind.

6. Start curing as soon as possible after finishing.

7. Keep concrete damp during the entire curing period.

C. Glossary

Admixture. A material other than water, aggregates, or cement, used as an ingredient of concrete or mortar, and added to the batch immediately before or during its mixing.

Aggregate. Granular material such as natural sand, manufactured sand, gravel, crushed gravel, crushed stone, and air-cooled iron blast-furnace slag, which when bound together into a conglomerated mass by a paste forms concrete or mortar.

Air content. The volume of air voids in cement paste, mortar, or concrete, not including pore space in aggregate particles; usually expressed as a percentage of total volume of the paste, mortar, or concrete.

Air-entraining agent. An addition for hydraulic cement or an admixture for concrete or mortar, it causes air to be incorporated as tiny bubbles in the concrete or mortar during mixing, usually to increase its workability and frost resistance.

Air-entraining cement. Hydraulic cement containing an air-entraining addition in amounts that cause the product to entrain air in mortar up to certain limits.

Air entrainment. The production of air in the form of tiny bubbles (generally smaller than 1 mm) during the mixing of concrete or mortar.

Anchor bolt. A bolt with the threaded portion projecting from a structure, generally used to hold the frame of a building secure against wind load.

Bar support (also *Bar chair*). A rigid device used to support and/or hold reinforcing bars in proper position to prevent displacement before or during concreting.

Batter. The gradual slope of a wall from bottom to top.

Bleeding. The flow of mixing water within, or its emergence from, newly placed concrete or mortar.

Block. A concrete masonry unit, usually containing hollow cores.

Bond. Adhesion and grip of concrete or mortar to reinforcement or to other surfaces against which it is placed, including friction due to shrinkage and longitudinal shear in the concrete engaged by the bar deformations; the adhesion of cement paste to aggregate.

Broom finish. Surface texture resulting from stroking a broom over freshly placed concrete.

Bug holes. Small regular or irregular cavities, usually not exceeding 15 mm in diameter, caused by entrapment of air bubbles in the surface of formed concrete during placing and compaction.

Cast-in-place. Mortar or concrete deposited in the place where it must harden as part of the structure (as opposed to precast concrete).

Cement, hydraulic. A cement capable of setting and hardening under water.

Cement, masonry. Hydraulic cement produced for use in mortars for masonry construction where greater plasticity and water retention are desired than can be obtained using portland cement alone.

Cement, portland. Product obtained by pulverizing clinker consisting essentially of hydraulic calcium silicates; usually containing calcium sulfates as an interground addition.

Cement, white. Portland cement which hydrates to a white paste; made from raw materials of low iron content.

Cement content. Quantity of cement contained in a unit volume of concrete or mortar, preferably expressed as weight.

Cement paste. A mixture of cement and water; may be either hardened or unhardened.

Coarse aggregate. Larger aggregate in a mix, such as gravel or crushed stone, usually from ¼ inch diameter up.

Compressive strength. The measured maximum resistance of a concrete or mortar sample to axial loading; expressed as force per unit cross-sectional area; expressed in pounds per square inch.

Concrete, green. Concrete that has set but not appreciably hardened.

Consistency. The ability of freshly mixed concrete or mortar to flow; the usual measurements for concrete is slump.

Control joint. Formed, sawed, or tooled groove in a concrete structure that regulates the location and amount of cracking and separation due to dimensional change of different parts of a structure; avoids development of high stresses.

Cover. In reinforced concrete, the least distance between the surface of the reinforcement and the entire surface of the concrete.

Crazing cracks. Fine, random cracks or fissures caused by shrinkage; can appear in a surface of cement paste, mortar, or concrete.

Curing. Maintenance of humidity and temperature of

freshly placed concrete during some definite period following placing, casting or finishing, to assure satisfactory hydration and proper hardening of the concrete.

Dry pack. To forcibly ram a moist portland cement aggregate mixture into a confined area; also the mixture so placed.

Durability. The ability of concrete to resist weathering action, chemical attack, abrasion, and other conditions of service.

Early strength. Strength of concrete or mortar developed soon after placement, usually during the first 72 hours.

Efflorescence. A deposit of salts—usually white— emerging from below and forming on the surface.

Epoxy resins. A class of chemical bonding systems used in preparation of special coatings or adhesives for concrete, or as binders in epoxy-resin mortars and concrete.

Expansion joint. Also called an isolation joint. A separation between adjoining parts of a concrete structure, provided to allow small relative movements (such as those caused by thermal changes) to occur independent of each other and without serious damage.

Exposed-aggregate finish. A decorative finish for concrete work achieved by removing, generally before the concrete has fully hardened, the outer skin of mortar and exposing the coarse aggregate.

Fine aggregate. Smaller aggregate in a mix, such as sand.

Finishing. Leveling, smoothing, compacting, and otherwise treating surfaces of fresh or recently placed concrete or mortar.

Floating. The operation of finishing a fresh concrete or mortar surface by use of a float.

Footing. That portion of the foundation of a structure which spreads and transmits loads directly to the soil.

Formwork. Total system of support for freshly placed concrete including the mold or sheathing that contacts the concrete, as well as all supporting members, hardware, and necessary bracing.

Honeycomb. Voids left in concrete due to failure of the mortar to effectively fill the spaces among coarse aggregate particles.

Hydration. In concrete, the chemical reaction between cement and water.

Masonry. Construction composed of shaped or molded units, usually small enough to be handled by one man, such as brick or block.

Masonry mortar. Mortar used in masonry structures.

Mold. A form used in the fabrication of precast mortar or concrete units.

Mortar. A mixture of cement paste and sand.

Oversanded. Containing more sand than necessary for adequate workability and satisfactory finishing.

Popout. Small portions of a concrete surface that break away due to internal pressure and leave a shallow, typically conical, depression.

Precast concrete. Concrete cast elsewhere than its final position in the structure.

Proportioning. Selection of proportions of ingredients for mortar or concrete that give the most economical use of available materials to produce mortar or concrete of required properties.

Ready-mixed concrete. Concrete manufactured for delivery to a purchaser while in a plastic, unhardened state.

Reinforced concrete. Concrete containing reinforcement and designed on the assumption that the two materials act together to resist forces.

Reinforcement. Metal bars, wires, or other slender members embedded in concrete so that the metal and the concrete act together in resisting forces.

Reinforcement, mesh. An arrangement of bars or wire, normally in two directions at right angles, tied or welded at the intersections.

Retarder. An admixture which delays the setting of cement paste, and hence of mixtures such as mortar or concrete.

Scaling. Local flaking or peeling away of the near-surface portion of concrete or mortar.

Set. The condition reached by a cement paste, mortar, or concrete when it has lost plasticity to a particular degree, usually measured in terms of resistance to penetration or deformation; initial set refers to first stiffening; final set refers to attainment of significant rigidity.

Shrinkage. Volume decrease caused by drying and chemical changes.

Slump. A measure of consistency of freshly mixed concrete.

Spall. A fragment, usually in the shape of a flake, detached from a larger mass by a blow, by the action of weather, by pressure, or by expansion within the larger mass.

Tensile strength. The greatest longitudinal stress a material can resist without tearing apart.

Undersanded. With respect to concrete: containing too low a proportion of fine aggregate for best properties in the fresh mixture, especially workability and finishing characteristics.

Water-cement ratio. The ratio of the amount of water, not including that absorbed by the aggregates, to the amount of cement in a concrete or mortar mixture; preferably stated as a decimal, by weight.

Welded-wire fabric reinforcement. Welded-wire fabric in either sheets or rolls, used to reinforce concrete.

D. CONVERSION FACTORS
U.S. Customary to SI (Metric)*

To convert from	To	Multiply By
Length		
foot (ft)	meter (m)	0.3048 E**
inch (in)	centimeter (cm)	2.54 E
yard (yd)	meter (m)	0.9144 E
Area		
square foot (sq ft)	square meter (m²)	0.0929
square inch (sq in)	square centimeter (cm²)	6.451
square yard (sq yd)	square meter (m²)	0.8361
Volume (capacity)		
cubic foot (cu ft)	cubic meter (m³)	0.02832
gallon (gal)	cubic meter (m³)	0.003785
ounce (oz)	cubic centimeter (cm³)	29.57
Pressure or Stress (force per area)		
pound-force/ square foot (psf)	kilogram-force/ square meter (kgf/m²)	4.882
pound-force/ square foot (psf)	newton/square meter (N/m²)	47.88
pound-force/ square inch (psi)	kilogram-force/ square centimeter (kgf/cm²)	0.07031
pound-force/ square inch (psi)	newton/square meter (N/m²)	6895
Mass		
ounce-mass (avdp)	gram (g)	28.35
pound-mass (avdp)	kilogram (kg)	0.4536
ton (metric)	kilogram (kg)	1000 E
ton (short, 2000 lbm)	kilogram (kg)	907.2
Mass per Volume		
pound-mass/ cubic foot (pcf)	kilogram/cubic meter (kg/m³)	16.02
pound-mass/ cubic yard (pcy)	kilogram/cubic meter (kg/m³)	0.5933

*Selected list of units commonly used in concrete technology. The reference source for information on SI units and more exact conversion factors is "Metric Practice Guide" ASTM E 380, American Society for Testing and Materials.

See also *American Metric Construction Handbook,* by R. J. Lytle, 1976, Structures Publishing Company.

** E indicates that the factor given is exact.

E. Product Information

An encyclopedia could be prepared to list all the products available for use with concrete, and a second book listing types of concrete products and where to buy them. Some items may be available nationwide, others are manufactured locally under franchise agreements and may not be available in your area.

When you are looking for a contractor, supplier, or equipment and materials, refer to the classified advertising section of your local newspaper and the classified telephone directory. Some of the classifications that apply are: Brick, block and cement; Building materials; Cement; Concrete additivies; Concrete aggregates; Concrete blocks and shapes; Concrete contractors; Concrete mixers; Concrete products; Concrete: transix-mixed; Landscape contractors; Landscaping supplies; Mason contractors; Patio builders; Patio materials; Rentals: equipment; Stone.

You may also contact industry and technical organizations for information and to locate a supplier or a specialist in your locality. In addition to national trade associations, there are state and local associations. Here are a few national organizations; several publish bulletins and handbooks for sale.

American Concrete Institute, P.O. Box 19150, Detroit, Michigan 48219 (publications for sale).

American Society for Concrete Construction, 329 Interstate Road, Addison, Illinois 60101.

Mason Contractors of America, 208 South LaSalle Street, Room 480, Chicago, Illinois 60601.

National Concrete Masonry Association, P.O. Box 135, McLean, Virginia 22201.

National Precast Concrete Association, 825 East 64th Street, Indianapolis, Indiana 46220.

National Ready Mixed Concrete Association, 900 Spring Street, Silver Spring, Maryland 20910.

Portland Cement Association, 5420 Old Orchard Road, Skokie, Illinois 60076 (publications for sale).

For more information on some of the products mentioned in this book, or to locate a local distributor or producer, try contacting these sources.

Borden Chemical, 1500 Touhy Avenue, Elk Grove Village, Illinois 60007—sealants and bonding agents.

Decor Molds, 1083 Bloomfield Avenue, West Caldwell, New Jersey 07006—Rollamix concrete mixer and plastic molds for patio block and screen block.

Form Incorporated, P.O. Box K, South Lyon, Michigan 48178—concrete playground sculptures.

Goldblatt Tool Company, 511 Osage, Kansas City, Kansas 66110—cement mason's and bricklaying tools.

Grass Pavers Limited, 3807 Crooks Road, Royal Oak, Michigan 48073—Monoslabs/Grass Pavers precast turf-grids.

KNR Concrete Systems Ltd., 1655 Sismet Road, Unit 20, Mississauga, Ontario L4W1Z4, Canada—Knauer paving stones.

Lotel, Inc., P.O. Box 14624, Baton Rouge, Louisiana 70808—plastic bar chairs to support reinforcement.

Master Builders, Lee at Mayfield, Cleveland, Ohio 44118—admixtures, hardeners, and toppings.

W. R. Meadows, Inc., 2 Kimball Street, Elgin, Illinois 60120—admixtures, bonding agents, coatings, joint sealants.

PCA Industries, Inc., 29-24 40th Avenue, Long Island City, New York 11101—concrete playground sculptures.

P.E.E.R. Institute, 76 Brookton, Highway, Klemscott, Western Australia 6111, Australia—Sculpcrete process.

Reichard-Coulston, Inc., 15 East 26th Street, New York, New York 10010—natural and synthetic pigments for coloring concrete.

L. M. Scofield Company, 5511 East Slauson Avenue, Los Angeles, California 90040—color agents and coatings for concrete.

Sika Chemical Corporation—Box 297, Lyndhurst, New Jersey 07071—admixtures, coatings, adhesives, joint sealants, and patching compounds.

Sta-Crete, Inc., 893 Folsom Street, San Francisco, California 94107—coatings, toppings, adhesives, sealants, patching compounds.

Standard Dry Wall Products, 7800 N.W. 38th Street, Miami, Florida 33166—waterproofing, restoration, protective products and treatments for masonry and concrete.

Index

Other SUCCESSFUL Books

SUCCESSFUL SPACE SAVING AT HOME. The conquest of inner space in apartments, whether tiny or ample, and homes, inside and out. Storage and built-in possibilities for all living areas, with a special section of illustrated tips from the professional space planners. 8½″ x 11″; 128 pp; over 150 B-W and color photographs and illustrations. $12.00 Cloth. $4.95 Paper.

BOOK OF SUCCESSFUL HOME PLANS. Published in cooperation with Home Planners, Inc.; designs by Richard B. Pollman. A collection of 226 outstanding home plans, plus information on standards and clearances as outlined in HUD's *Manual of Acceptable Practices.* 8½″ x 11″; 192 pp; over 500 illustrations. $12.00 Cloth. $4.95 Paper.

FINDING & FIXING THE OLDER HOME, Schram. Tells how to check for tell-tale signs of damage when looking for homes and how to appraise and finance them. Points out the particular problems found in older homes, with instructions on how to remedy them. 8½″ x 11″; 160 pp; over 200 photographs and illustrations. $12.00 Cloth. $4.95 Paper.

WALL COVERINGS AND DECORATION, Banov. Describes and evaluates different types of papers, fabrics, foils and vinyls, and paneling. Chapters on art selection, principles of design and color. Complete installation instructions for all materials. 8½″ x 11″; 136 pp; over 150 B-W and color photographs and illustrations. $12.00 Cloth. $4.95 Paper.

BOOK OF SUCCESSFUL FIREPLACES, Lytle. How to build, decorate, and use all types of fireplaces. Covers fireplace construction, history, problems, cookery, even how to keep a good fire going. 8½″ x 11″; 104 pp; over 150 B-W and color photographs and illustrations. (Chosen by Popular Science Book Club). $12.00 Cloth. $4.95 Paper.

BOOK OF SUCCESSFUL KITCHENS, Galvin. In-depth information on building, decorating, modernizing, and using kitchens, by the editor of *Kitchen Business* magazine. 8½″ x 11″; 136 pp; over 200 B-W and color photographs and illustrations. $12.00 Cloth. $4.95 Paper.

BOOK OF SUCCESSFUL PAINTING, Banov. Everything about painting any surface, inside or outside. Includes surface preparation, paint selection and application, problems, and color in decorating. "Before dipping brush into paint, a few hours spent with this authoritative guide could head off disaster." —*Publishers Weekly.* 8½″ x 11″; 114 pp; over 150 B-W and color photographs and illustrations. $12.00 Cloth. $4.95 Paper.

BOOK OF SUCCESSFUL BATHROOMS, Schram. Complete guide to remodeling or decorating a bathroom to suit individual needs and tastes. Materials are recommended that have more than one function, need no periodic refinishing, and fit into different budgets. Complete installation instructions. 8½″ x 11″; 128 pp; over 200 B-W and color photographs. (Chosen by Interior Design, Woman's How-to, and Popular Science Book Clubs). $12.00 Cloth. $4.95 Paper.

TOTAL HOME PROTECTION, Miller. How to make your home burglarproof, fireproof, accidentproof, termiteproof, windproof, and lightningproof. With specific instructions and product recommendations. 8½″ x 11″; 124 pp; over 150 photographs and illustrations. (Chosen by McGraw-Hill's Architects Book Club). $12.00 Cloth. $4.95 Paper.

BOOK OF SUCCESSFUL SWIMMING POOLS, Derven and Nichols. Everything the present or would-be pool owner should know, from what kind of pool he can afford and site location, to construction, energy savings, accessories and maintenance and safety. 8½″ x 11″; 128 pp; over 250 B-W and color photographs and illustrations. $12.00 Cloth. $4.95 Paper.

HOW TO BUILD YOUR OWN HOME, Reschke. Construction methods and instructions for wood-frame ranch, one-and-a-half story, two-story, and split level homes, with specific recommendations for materials and products. 8½″ x 11″; 336 pp; over 600 photographs, illustrations, and charts. (Main selection for McGraw-Hill's Engineers Book Club). $14.00 Cloth. $5.95 Paper.

HOW TO CUT YOUR ENERGY BILLS, Derven and Nichols. A homeowner's guide designed not for just the fix-it person, but for everyone. Instructions on how to save money and fuel in all areas—lighting, appliances, insulation, caulking, and much more. If it's on your utility bill, you'll find it here. 8½″ x 11″; 136 pp; over 200 photographs and illustrations. $12.00 Cloth. $4.95 Paper.

Structures Publishing Company   Box 423   Farmington, Michigan 48024